Porous Media: Heat and Mass Transfer, Transport and Mechanics

Porous Media: Heat and Mass Transfer, Transport and Mechanics

José Luis Acosta
and
Andrés Felipe Camacho
Editors

Nova Science Publishers, Inc.
New York

For permission to use material from this book please contact us:
Telephone 631-231-7269; Fax 631-231-8175
Web Site: http://www.novapublishers.com

LIBRARY OF CONGRESS CATALOGING-IN-PUBLICATION DATA

Acosta, Jose Luis.
Porous media : heat & mass transfer, transport & mechanics / Jose Luis Acosta & Andres Felipe Camacho.
p. cm.
Includes bibliographical references and index.
ISBN 978-1-60692-437-2 (hardcover)
1. Porous materials--Industrial applications. 2. Porous materials--Mechanical properties. 3. Diffusion--Mathematical models. 4. Heat--Transmission--Mathematical models. I. Camacho, Andres Felipe. II. Title.
TA418.9.P6A27 2009
620.1'169--dc22
2008045771

Published by Nova Science Publishers, Inc. ✦ New York

CONTENTS

PREFACE

Chapter 1 - Most research on porous media is based on a continuum model that analyze transport processes in a macroscale approximation, averaging on a large number of pores, neglecting poral scale effects.

In reactive solute flow dynamics this approach can produce misleading results, because reactants that are considered homogenized at Darcy scale are not actually perfectly mixed. Numerical schemes based on this assumption are unable to reproduce properly experimental results, because they are disregarding inhomogeneities at poral scale, just where reactions are taking place.

This lack of homogeneity at poral scale (called segregation effect) can be addressed in the transport equation using an effective reaction rate term, acting as a source. Within this approach the main problem seems to be how to model the effective reaction rate. A phenomenological model is proposed and its parameter is estimated by minimizing the error between simulated results and experimental data. Correlation between the parameter of the mathematical model and physical properties of the experimental set up is discussed.

Given that this kind of processes are of advective - diffusive - reactive type, they present three different characteristic time scales, depending on which is the dominant process. Solving this problem numerically, attention must be paid to this fact, in order to address the full complexity of this equation.

Chapter 2 - Stress changes induced by pore pressure or temperature changes occurring during fluid production (or injection) from (or into) porous reservoirs may lead to hazards such as ground movement, fluid leakage and earthquakes. Published data compiled for several reservoirs indicate that horizontal stress changes between 0.24 and 0.84 times the magnitude of the pore pressure change have been measured. Consequently, there is a clear need for models that predict pore pressure-induced stress changes. Temperature-induced stress changes, although less well documented, also have the potential to have significant effects. A review of three semi-analytical solutions for thermo-poroelasticity field equations is conducted in this chapter, covering: (1) the theory of strain nuclei; (2) the theory of inclusions; and (3) the theory of inhomogeneities. Closed-form solutions – most of which have been derived by the authors – are presented for pore pressure and thermally-induced stress changes within various geometrical variations of an ellipsoidal reservoir in a full space, for conditions where the mechanical properties of the reservoir are either identical to or different from the surrounding rock. Results for these models are presented in terms of dimensionless parameters, which facilitates their application to a broad range of reservoir conditions. A case study for the Ekofisk oil field, in the Norwegian sector of the North Sea, is used to demonstrate the use of these models. The results of this case study demonstrate that poroelastic modeling can predict horizontal stress changes with reasonable accuracy, and further challenge the common assumption that vertical stress changes are generally negligible.

Chapter 3 – Porous materials are inorganic or organic crosslinked or uncrosslinked contain pores of almost all sizes. The chapter is focused only on porous hydrophilic gels (hydrogels) with pore size from units to hundreds of micrometers, which are generally called macroporous or superporous. The macroporous hydrogels are most often used in the biomedical field, e.g. for tissue engineering, cell therapy and as biocompatible implants.

Porosity is usually achieved by one of the four methods:

Crosslinking polymerization in the presence of substances that are solvents for the monomers, but precipitants for the formed polymer.

Crosslinking polymerization in the presence of water soluble substances (sugars, salts) which are washed out from the hydrogel after polymerization.

Crosslinking polymerization in the presence of substances relasing gases which remain in the resulting hydrogel.

Freeze-sublimation of the hydrogel swollen in water.

By using any of the procedures, hydrogels with communicating or non-communicating pores can be obtained, i.e., with the pores that are or are not interconnected. This chapter is focused on the second-mentioned method, which makes it possible to adjust the pore size in a wide range and to prepare materials with communicating and non-communicating pores. Other methods allow control the pore size only in a small range; however, they are useful for some purposes as well. For most applications communicating pores are preferred.

Chapter 4 - How fast can drug molecules escape from a controlled matrix-type release system? This important question is of both, scientific and practical importance, as increasing emphasis is placed on design considerations that can be addressed only if the physical chemistry of drug release is better understood. In this work, the labor performed by our group on the modeling of diffusion-controlled release systems is reviewed and updated with the inclusion of new results.

Chapter 5 - Surfactant-enhanced aquifer remediation (SEAR) is an excellent source-depletion technology for the remediation of the subsurface contaminated by non-aqueous phase liquid (NAPL). Modeling of surfactant and contaminant transport in the porous media can provide quantitative assessment and prediction of SEAR performance. Processes including surfactant micellization, surfactant and contaminant sorption on the aquifer media, contaminant micellar-solubilization, contaminant adsolubilization by admicelles/hemimicelles formed on the aquifer media, advective and dispersive transport of the surfactant and contaminant, reduction of contaminant residual saturation, and alteration of interfacial properties may be involved in SEAR. The reaction kinetics and heterogeneities of the aquifer properties further complicate modeling of these processes. A number of models have been developed to describe the transport behavior of surfactant and contaminant, which include surfactant and contaminant coupled transport model, solute transport model coupled with two- or multi-site sorption, multi-component reactive transport model, multi-component multi-phase flow model, and UTCHEM model, etc. In this review, assumptions and reactive processes as well as the mathematical formulations of these models are reviewed. It is recommended that the processes involved in

SEAR be characterized independently with controlled experiments to examine their individual contribution to the overall contaminant fate and transport. Increased understanding of the fundamental hydrodynamic and physiochemical processes occurring during SEAR will significantly improve the accuracy of SEAR modeling.

Chapter 6 - Porous inorganic membranes offer profound potential in industrial applications for gas separation in a high temperature catalytic reaction. They are also potentially useful for the separation of trace contaminants such as ammonia, hydrogen sulfide and carbonyl sulfide from coal-gasifier fuel gases. A particularly important application is in fuel cell, in which, H_2 and O_2 are used as feed gas and the membrane acts as a solid electrolyte (and/or electrode) in order to generate electricity. Other promising applications with which porous inorganic membranes have received considerable attention in a wide array of industrial operations include gas permeation for separation of CO_2 or H_2 from methane and other hydrocarbons, adjustment of H_2/CO ratio in syngas, separation of air into nitrogen and oxygen, recovery of helium and methane from bio-gas. Gas permeation involves feeding of gas mixture usually at a higher pressure or concentration into feed stream. Lower molecular weight gas species is separated from higher molecular weight gas species by virtue of different rate of diffusion through the membrane. Lower molecular weight species, having higher rate of diffusion diffuse faster into a much lower pressure or concentration in the permeate stream of the membrane. Permeation of gas species in micro and meso porous inorganic membrane has been known to be contributed primarily by the two main transport mechanisms. The first being, Knudsen diffusion mechanism and second is surface diffusion. The discussion on the internal transport mechanism by the former is relatively ample in the past and present literatures, but the transport mechanism due to the latter, however, appear to be meager possibly due to the inherent difficulty and inadequacy of robust techniques to measure it. In this paper, the internal gas transport as a result of surface diffusion mechanism is given the necessary attention. The correlation between the surface mechanism and loading factor is also presented.

Chapter 7 - The porosity, tortuosity, permeability and heat exchange characteristics from binary packings, containing mixtures of small d and large D spherical particles, are analysed in the present work. Binary packing porosity (ε), tortuosity (τ), permeability and heat exchange performance are dependent on the volume fraction of large particles, x_D, present in the mixtures, as well as on the particle size ratio, $\delta = d/D$. In the region of minimum porosity from the binary mixtures (containing spheres with diameter d and D), heat exchange performance and permeability from binary packing are higher than that of the packing containing the small particles d alone (mono-size packing). The δ region where the permeability of binary packing is higher than the permeability of mono-size packing of particles d is located in the range $0.1 \leq \delta < 1.0$. An increase in permeability by a factor of two is achieved for particle size ratios between 0.3 and 0.5. Tortuosity can be modelled by the simple function $\tau = 1/\varepsilon^{\alpha}$ and it is shown that, in the region of minimum porosity, α varies between 0.5 (mono-size packing) and 0.4 (binary packing with δ close to 0.03). Due to the tortuosity increase, binary mixtures give rise to Kozeny's coefficients substantially higher than five.

Using the commercial finite element software package POLYFLOW® it was possible to confirm the heat exchange enhancement referred above. The obtained improvement on the thermal performance is related to the increase of effective thermal conductivity in the binary packing and to the increase in transversal thermal conductivity due to the porosity decrease and tortuosity increase. For non-Newtonian fluids from the power-law type, τ decreases with the decrease of the flow index behaviour

Chapter 8 - The transport mechanisms of colloids in saturated porous media have been studied in great details. Suspended particle (SP) transport in the subsurface has only recently attracted significant attention. This chapter presents an experimental study of the transport of suspended particles in a saturated porous media, aimed at delineating the effects of hydrodynamic and gravity forces on particle transport and deposition rate. Suspended particles were injected under saturated flow conditions into a laboratory column packed with gravel or glass beads. The measured particle breakthrough curves were well described by the analytical solution of an advection-dispersion equation with a first-order deposition kinetic. The laboratory tests performed with different flow rates and column materials showed that hydro-dispersive parameters derived from 1D advection-dispersion model depend on the flow rate, the suspended particles characteristics and the porous medium. The results provided the existence of a flow velocity beyond which particles travel faster than the conservative tracer. Hydraulically equivalent media with neighbouring porosities produce differences in the transport of suspended particles. The dispersion was found to be controlled by the pore-space geometry, the relative size of particles and the flow rate. This chapter shows that particles, pore sizes and flow rate are among the main mechanisms influencing the transport and deposition rate of SP in porous media.

Chapter 9 - To optimize the balance between physiological burden and protection, models were developed to describe the important processes at hand. By using these models, the optimum requirements for CBRN-protective clothing can be established. These requirements will lead to optimum values of the various properties of the clothing. A large variety of models can be identified, which are important for such a tool. A selection of models is presented here: Agent vapor breakthrough through the clothing, Air flow around, through and underneath the clothing, Agent vapor concentration underneath the clothing, Agent deposition onto the skin.

The vapor breakthrough model, the pressure distribution model and the deposition model were verified experimentally.

Short Communication A - In this study the influence of Soret and Dufour effects on the double diffusive natural convection induced by an heated isothermal wavy vertical surface in a fluid saturated porous enclosure under Darcian assumptions has been analysed. The mathematical model has been solved numerically by finite element method and the simulations are carried out for various values of parameters such as D_f (Dufour Number), S_r (Soret Number), Le (Lewis Number), B (Buoyancy Number) and N (Number of waves per unit length) at small values of Ra (Rayleigh Number).

In: Porous Media: Heat and Mass Transfer...
Editors: J. L. Acosta and A.F. Camacho

ISBN: 978-1-60692-437-2

Chapter 1

MODELING REACTIVE TRANSPORT DRIVEN BY SCALE DEPENDENT SEGREGATION

D. A. Cuch*[1]*, C. D. El Hasi*[1]*, D. Rubio*[2]*,
***G. C. Urcola*[1] *and A. Zalts*[1]**

[1] Instituto de Ciencias, Universidad Nacional de General Sarmiento
José M. Gutierrez 1150, B1613GSX - Los Polvorines,
Provincia de Buenos Aires - Argentina
[2] Centro de Matemática Aplicada, ECyT.
Universidad Nacional de General San Martín
25 de Mayo y Francia. CP 1650. San Martín,
Provincia de Buenos Aires, Argentina

ABSTRACT

Most research on porous media is based on a continuum model that analyze transport processes in a macroscale approximation, averaging on a large number of pores, neglecting poral scale effects.

In reactive solute flow dynamics this approach can produce misleading results, because reactants that are considered homogenized at Darcy scale are not actually perfectly mixed. Numerical schemes based on this assumption are unable to reproduce properly experimental results, because they are disregarding inhomogeneities at poral scale, just where reactions are taking place.

This lack of homogeneity at poral scale (called segregation effect) can be addressed in the transport equation using an effective reaction rate term, acting as a source. Within this approach the main problem seems to be how to model the effective reaction rate. A phenomenological model is proposed and its parameter is estimated by minimizing the error between simulated results and experimental data. Correlation between the parameter of the mathematical model and physical properties of the experimental set up is discussed.

Given that this kind of processes are of advective - diffusive - reactive type, they present three different characteristic time scales, depending on which is the dominant process. Solving this problem numerically, attention must be paid to this fact, in order to address the full complexity of this equation.

Key words: porous media, numerical methods, segregation, advection – reaction – diffusion equation, transport processes.

INTRODUCTION

All human activities take place in the earth's environment and lead to some kind of transaction with it. Human activities, from the most basic as our individual metabolism, to the most industrialized, are bounded by environmental conditions; on the other hand, they change the environment due resource consumption and waste emission into it. Human activities are substantially modifying the global carbon, nitrogen, sulfur and phosphorous cycles. Each of the four elements moves from one chemical state to another and from one physical location to another on the earth's surface. The cycles are powered by solar energy, in conjunction with the earth's gravity and geothermal energy. The nutrients, for example, flow among "reservoirs". The reservoirs of interest are life forms (living and dead plants and animals), the soil, the oceans and other water bodies, the atmosphere, and rocks (Ayres et al, 1997). They also move within each reservoir. Many environmental problems involve pollutant transport processes: there are sources and sinks, linked by the movement of these substances among the reservoirs. Air and water masses behave as mobile reservoirs, and are responsible of bulk advective pollutant transport. Analysis of solute flow dynamics takes into account that, in actual situations, there is also a spreading or mixing phenomenon associated to the advective movement: this dispersion spread out sharp fronts, resulting in the dilution of the solute. Interaction of the pollutants with the solid phase of soil or particulate matter, for example, through sorption processes results in retarded fronts and changes in concentration. Also, the transport of reacting species is affected by the changes produced in the chemical composition of the environment.

Considerable progress has been made in the study of reactive solute transport in heterogeneous groundwater (J. Kohne et al, 2006; Wriedt and Rode, 2006; Brusseau, 1994; Jardine et al., 1999, for example). Many different processes, acting concurrently, contribute to solute transport, generating difficulties to observe and quantify their behavior in the field. For example, the leaching of a substance into the soil is accompanied by other processes that modulate the transport due to interactions between the substance and soil components such as dispersion, retardation, degradation of chemicals, effects due to the heterogeneities of the soil, sorption, etc. (Logan, 1999; Ataie-Ashtiani, 2007). So, a modeling tool is needed to get a deeper understanding of the phenomena that can be applied to a large number of applications, such as waste disposal management, drinking water supply protection and environmental remediation.

Although it is possible to model processes such as water flow through a porous media using Darcy's law, and also the chemical and biological reactions taking place during transport in one, two and three dimensions, several aspects remain to be considered. Using a fluidistic approach, the concentration of solutes can be estimated solving the hydrodynamic equations taking into account parameters of an aquifer, for example, and suitable boundary conditions. Under the supposition of the homogeneity of the system, deterministic methods can be used to get approximated results. As a first approximation, these results cannot address appropriately the influence of inhomogeneities that affect hydraulic conductivity and dispersion. Stochastic models are more suitable in these cases.

An important characteristic of environmental systems is their spatial and temporal variability in chemical composition. As in many cases the main sources of substances released to the environment are heterogeneous in their spatial and temporal distribution (traffic or industry for air pollution; waste and sewage management, industrial facilities, and agricultural practices for soil and groundwater contamination, for example), the observational window (scale) used to model, leads to different results if the heterogeneities are preserved within the flow or not, and depending on how this is related to the overall chemistry.

In almost all cases dealing with transport in porous media, a mean concentration over several pores is considered. As a consequence, the system is described as a continuous media. If there is a concentration gradient, molecular diffusion acts at very short spatial range, following Fick's law. Moreover, solvent velocity is not always spatially homogeneous and the solute can be spreading along the flow path, in a process called mechanical dispersion. Both physical processes contribute to hydrodynamical dispersion causing the mixing and dilution of solutes.

A very important issue is the determination of dispersion coefficient (Gelhar, 1993). Some studies have shown that it is not possible to understand mixing and dilution only under the approach of mean concentrations (Cao and Kitanidis, 1998; Raje and Kapoor, 2000). Rashidi et al. (1996) found experimentally the dependency of solute concentration with the pore size, while Cao and Kitanidis (1998) did the same numerically. Both coincide that concentration at poral scale is ruled by molecular diffusivity. These non reactive transport studies have been extended to the analysis of first order decay cases with spatially variable decay rate. Processes with several species have been considered (Dykaar and Kitanidis, 1996). When these models are extended to reactive processes usually the reaction rate is obtained from batch experiments. When several species are involved, the reaction rate usually is not a linear function of their concentrations, producing misleading results. This observation has been emphasized for bimolecular reactions: theoretically by Kapoor et al. (1997) in the case of a laminar shear flow and a random porous media flow, experimentally in a Posseuille flow (Kapoor, 1998) and also experimentally by Raje and Kapoor (2000) in a packed column of glass beads. In these cases, ignoring correlation at microscale, the simulations predict a higher product concentration than the actual one.

SEGREGATION

Owing to an increasing interest in environmental problems, considerable attention has been focused on the prediction of concentration levels in flowing systems such as the dispersion of pollutants, reactive plumes, the effects of turbulent flows, and mixing and segregation rates.

When modeling transport and fate of environmentally important substances, the spatial and temporal variability in chemical composition must be taken into account. In reactive systems, the heterogeneities influence and modulate the chemistry. So, the understanding of the interaction between chemistry and the dynamics can be scale related. For example, for many species of interest in urban atmospheric pollution, having reaction time scales that are longer than the travel time across the urban area, chemical reactions can be ignored in describing local dispersion from strong individual sources (Stein et al, 2007). Instead, when updraft plumes in the convective boundary layer are carrying the pollutants from the surface and distributing them within this layer in a few minutes, a time scale quite similar to that of the ozone chemistry, important interactions between turbulence and ozone chemistry can be expected (Auger and Legras, 2007).

Segregation in Urban Atmospheric Modeling

Our knowledge about atmospheric phenomena derives largely from observations. In dealing with atmospheric data, one must make allowances for the mobility of the atmosphere and a tendency of all measurable quantities to undergo sizable fluctuations. These variations arise in part from temporal changes in the production mechanisms (sources) and the removal processes (sinks). The irregularities of air motions responsible for the spreading of trace substances within the atmosphere impose additional random fluctuations on local concentrations. In addition to random fluctuations, atmospheric data often reveal periodic variability, usually in the form of diurnal or seasonal cycles. The proper choice of averaging period obviously becomes important for bringing out suspected periodicities (Warneck, 2000).

A common assumption of most atmospheric models is that all the pollutants released at the surface are rapidly distributed by convection within the boundary layer where they mix and react with other species. But an heterogeneous distribution of reactants and multiple or chain reactions in atmospheric chemistry are common: they will either accelerate or retard the reaction rates, so special attention must be paid to situations involving trace amounts of reactants in chemical reactions whose characteristic reaction times are much less than the atmosphere´s characteristic mixing time. In these cases segregation of the reactants can be severe, and may lead to long – term, volume averaged reaction rates which are significantly different from well-mixed reactions (Hilst, 1998).

A general property, first noticed by Galmarini et al. (1995) for a convective plume case, is that fast and reactive species combine to produce negative segregation. When the emissions are not uniform, it has been shown that segregation can reach significantly large negative values (Molemaker and Vilà-Guerau de Arellano, 1998). In these studies, based on a simple $A + B \rightarrow C$ chemical scheme, the species B is initially distributed within the domain and the species A is emitted at the top of the boundary. Negative segregation results from the fact that a local positive fluctuation in A leaves less B after the reaction occurred. Another possibility that leads to negative segregation is that surface emissions are separated (some species emitted on one part of the surface at the ground, the other on the rest of the surface). It depends on the grid size and approximation level if spatial separation among pollutants in emission is considered or not.

Auger and Legras (2007) found that if pollutants are emitted with uniform composition over the domain, although not necessarily uniformly, this composition is preserved by transport and diffusion which are linear processes in tracer concentration; segregation will be positive. If chemistry is slow with respect to dynamics, segregation can still be positive. In their case, the negative values can only be explained by fast chemical reactions, in particular those involving chemical species not present within the emission but created within the turbulent boundary layer. These authors show that all the compounds with characteristic time larger than one day behave like passive tracers and get mixed before they could react significantly. Instead, species with large Damköhler numbers are not affected by dynamics.

In most cases segregation is of the order or less than 1% when emissions are homogeneous at the ground. Although at a first glance it looks that segregation effects are negligible, nonlinearities in the chemistry are responsible that a 1% segregation leads to much larger effects on the mean concentrations. When heterogeneity is added, the dynamic exchanges between areas of different concentration considerably enhance the fluctuations and lead to much larger segregation than expected (Auger and Legras, 2007).

The worth of the more precise models depends on the objective of the model simulation. If the objective of the model is to mimic the reaction system in situ and the chemistry is influenced by initial or induced reactant segregation and/or by diffusion limited situation, the more complete model is required. On the other hand, if only a long – term global average is of interest, much cruder models will suffice. (Hilst, 1998).

REACTIVE TRANSPORT AND FLOW MODELING

Recently, the number of studies in environmental issues, related to reactive transport has increased, showing the importance of this aspect (Lajer Højberg et al, 2005; Jørgansen et al, 2004; Stutter et al, 2005; Steefel et al., 2005; Wriedt and Rode, 2006; Köhne et al, 2006, for example). Soil is a very complex porous medium, formed by a solid phase or matrix and poral voids that can be occupied by one or more fluid phases: liquid, usually water, and/or air. When water occupies all the pores, the soil is saturated.

Simplified conceptual models of soil processes are very useful to predict and / or understand the movement of different species in the environment. An important number of the models are deterministic, based on conservation laws for mass, energy and momentum. Usually, it is needed to understand how these models work to solve partial differential equations. Subsurface transport processes are ruled by Darcy's law and conservation of mass. The main objective is to calculate concentration of chemicals dissolved in water in space and time (Bear, 1988). Changes in chemical concentration are mainly due to: i) advective transport of the solutes with the flow, ii) hydrodynamic dispersion, given by diffusion or differential flow due to small scale changes in velocity, iii) the existence of sources or sinks, including chemical reactions, sorption, biodegradation as well as physical sources or sinks.

An interesting review of some of the major aspects dealing with transport of reactive contaminants, but restricted to just only one species, in heterogeneous subsurface environments is that of Brusseau (Brusseau, 1994). Several field experimental results are analyzed and the conceptual and mathematical approaches used are discussed. There are also several examples in the literature dealing with the transport equation with a single species under several circumstances (Valocchi and Malmstead, 1992; Ataie-Ashtiani and Hosseini, 2005; Kamra and Lennartz, 2005).

When analyzing the transport of solutes in porous media to understand the dilution, mixing and reaction processes it is possible to recognize several scales: i) a molecular scale where reactive processes are ruled by the shape or orientation of molecules and presence of catalysers, for example, ii) a continuum scale in the sense of fluid mechanics, where we must take into account interactions between dissolved substances in water and interactions with the solid matrix; this is the so called microscale, where each drop of fluid contains a large number of molecules, iii) a larger scale (Darcy's scale), where the Representative Elementary Volume (REV) contains a large number of pores, in such a way that the constitutive properties of the continuous media can be determined in laboratory experiments using macroscopic concepts such as porosity, dispersivity and hydraulic conductivity, between others.

In most cases, the models of solute transport in porous media in the literature are based on a single species, whose concentration is given as an averaged value over many pore spaces. So, the porous media is considered a continuous phase (Bear, 1988). When flow processes with multiple reactive species are studied, it is important to consider that the reactions take place at a molecular scale, which is much smaller than the poral scale. In this range the fluid velocity is never homogeneous in space, and the continuum hypothesis loses validity. Some authors (Kapoor et al., 1997; Kapoor et al., 1998, Raje and Kapoor, 2000; Gramling et al., 2002) discuss what happened in reactive transport emphasizing the differences between continuous models and actual measurements.

Although long term field concentration data are desirable to understand reactive transport in the environment, they are costly and time consuming. Usually, a combination of laboratory based experiments and numerical simulations and modeling is used to get a better understanding of the mechanisms and phenomena involved in these processes. In many cases, a suitable approach is to analyze simplified schemes, such as considering one-dimensional flow linked to bimolecular reactions (Raje and Kapoor, 2000; Gramling et al., 2002). The differences between empirical and numerical data depend on how these processes are modeled. While the equations are based on a continuum hypothesis, averaging an elevated pore number, the reaction dynamics are governed by poral scale processes (Kapoor et al., 1997). The parameters of the simulations are determined in batch reactors and supposed to be valid in transport processes. Recent mathematical models that describe diffusive-reactive processes more accurately (Meile and Tuncay, 2006) have been developed.

The Transport Equation

Transport in porous media includes solid and fluid phases, which can be liquid and/or gaseous. Its description is more complex than transport in just one phase, as for example substances dissolved in surface water or the movement of gases in the atmosphere. To describe transport in porous media we have to specify the relationship between solid and solvent (porosity), flux velocity, dispersion – diffusion coefficient, interface processes (adsorption) and reactive rate.

Solid, liquid and gas volumes (V_s, V_l and V_g, respectively), are related to the total porous medium volume V:

$$V = V_g + V_l + V_s \,, \tag{1}$$

We can rewrite the equation

$$1=\frac{V_g}{V}+\frac{V_l}{V}+\frac{V_s}{V}\equiv a+\theta+\frac{V_s}{V}, \qquad (2)$$

where a and θ are the volumetric content of air and water, respectively. The porosity or void volume is:

$$\phi=a+\theta=1-\frac{V_s}{V}. \qquad (3)$$

For processes in the saturated zone, volumetric water content and porosity are the same: $\phi=\theta$.

Approximations and simplifications (models) are needed for mathematical analysis of the phenomena when dealing with complex systems such as porous media (Huyakorn and Pinder, 1983). The model is a tool to understand observations at macroscopic level of processes that take place at a microscopic scale.

So, the solid phase of a porous medium can be described as a rigid matrix, formed by a spatial net of channels and tubes, randomly connected, with variable length, section and orientation. As a result, at a larger scale, the channels have a more or less uniform distribution. The fluid phase moves with a laminar flow, so each channel can be described with a fixed flow, although this is not necessarily the case for the flux direction along the channels. The fluid is supposed to be Newtonian, incompressible and chemically inactive. Isothermal processes are generally considered.

The *Representative Elementary Volume (REV)* is an element small enough to consider that the properties of the porous medium are constant within it, but large enough to be sure that the porous medium is still present and not only a part of it (the solid matrix or the fluid phase). The path of the fluid phase is not straight, the fluid has to move around the solid particles and through pores of different sizes. A tortuosity factor, that takes into account the connectivity between the pores, is useful to describe this movement. This tortuosity factor affects strongly the hydraulic conductivity.

The differential equation that describes transport processes can be obtained considering the balance of stored, flowing or created solute in a given volume (Jury et al., 1991):

$$\frac{\partial C}{\partial t}+\nabla\cdot\mathbf{J}=R. \qquad (4)$$

This is a continuity equation, where C is the total concentration of solute (mass of solute by volume of soil, or $C=\theta\cdot c$, where c is the fraction of solute in the liquid phase), $\mathbf{J}$ the total flux of solute (mass flowing by unit area and by unit time) and R a source (or sink) term. The concentration of solute at a given point is given by the balance between the inlet and outlet solute fluxes and the existence of reactions.

The flux of dissolved solutes has two terms: the convection of solute moving with the solution and a diffusive flux due to the movement at molecular level.

The convective term can be modeled as $\mathbf{J}_w c$, being $\mathbf{J}_w$ the water flux, and c the concentration in liquid phase. But in a porous media $\mathbf{J}_w$ is an average quantity over several pores, so it does not represent an actual water flow. Water flux does not describe the movement through tortuous path, local movements or differential fluxes respect to the mean flux values. This extra movement is called mechanical dispersion, so we can write:

$$\mathbf{J} = \mathbf{J}_w c + \mathbf{J}_m, \tag{5}$$

where $\mathbf{J}_m$ is the mechanical dispersion. It depends on variations in the velocity profile through the saturated pores, variations in size of pores and tortuosities, ramifications and/or entanglements of the pores.

The movement of solute in porous media due to mechanical dispersion $\mathbf{J}_m$ is very complex. A simple model frequently used assumed that:

- The medium is homogeneous through the volume where the solute is transported,
- The mixing time of the solute by means of dispersion or transversal diffusion (in stream tubes of different velocities in a transverse direction respect to the main direction of advection) is short compared with the average advection time (Taylor, 1953).

In a simplified flux problem in a cylindrical tube, the second condition assumes that the mixing is complete before the solution exits the tube. Under these assumptions the dispersive flux follows a Fickian law, i.e. the driving force is the concentration gradient,

$$\mathbf{J}_m = -\mathbf{D}_m \cdot \nabla c, \tag{6}$$

where $\mathbf{D}_m$ is the tensor of mechanical dispersion. Using a suitable coordinate system the dispersion tensor can be diagonalized, being its principal values the longitudinal (along the flow direction, D_{long}) and transversal (D_{trans}) dispersion coefficients. These dispersion coefficients are proportional to the seepage (poral) velocity $\mathbf{V} = \mathbf{J}_w / \theta$, being θ the water content (or porosity) for a saturated medium:

$$D_{\text{long or trans}} = \lambda_{\text{long or trans}} |\mathbf{V}|, \tag{7}$$

where $\lambda_{\text{long or trans}}$ (in cm) is the longitudinal or transversal dispersivity, respectively (Scheidegger, 1954). The value of λ_{long}, the predominant coefficient, depends on the scale over which the average of water fluxes and concentrations are taken. Typical values ranges from 0,1 – 2 cm for laboratory packed columns and from 5 – 20 cm for field problems.

The other term that contributes to transport is diffusion at molecular level which for low flux rates it could be the most important term:

$$\mathbf{J}_d = -\mathbf{D} \cdot \nabla c \,, \tag{8}$$

where $\mathbf{D}$ is the diffusion tensor in soil. From experimental measurements it is known that typical values for the principal diffusion coefficients are around $10^{-6}\,\mathrm{cm}^2/\mathrm{s}$.

Under these assumptions the model leads to a total flux term:

$$\mathbf{J} = \mathbf{J}_w c - \mathbf{D}_m \cdot \nabla c - \mathbf{D} \cdot \nabla c \equiv \mathbf{J}_w c - \mathbf{D}_e \cdot \nabla c \,, \tag{9}$$

where $\mathbf{D}_e$ is the effective dispersion – diffusion tensor. Normally $\mathbf{D}_e$ is dominated by the mechanical – dispersion term if fluxes are not too slow.

Under these assumptions the transport equation becomes:

$$\frac{\partial(\theta \cdot c)}{\partial t} + \nabla \cdot (\mathbf{J}_w c) - \nabla \cdot (\mathbf{D}_e \cdot \nabla c) = R \,. \tag{10}$$

This is an equation of advective – diffusive – reactive type. In a general case, numerical schemes are needed to solve it and some assumptions must be done in order to obtain a more suitable form of the equation. The simplest case would be a non reactive solute: $R = 0$, with negligible adsorption. Tracers as chloride or bromide ions dissolved in water and flowing through a saturated porous media with a neutral or negative charged matrix can be a good example of this situation.

The equation describing transport of solute is of second order and can be classified according to its mathematical properties. There are three types of second order differential equations: parabolic (as the one of thermal diffusion), elliptic (as in the potential theory, or boundary value problems) and hyperbolic (as the wave equation). Each of these equations has their own characteristics that must be taken into account to solve them. The mathematical complexity of the equation that governs the advective – dispersive – reactive type processes depends on which is the dominant term. Its mathematical structure must be considered carefully when solving the problem, and the method to be used has to be selected specifically in each case. Only in some particular cases it is possible to solve this equation exactly (Cokca, 2003). In a more general situation, several numerical techniques are required (Wheeler and Dawson, 1987; Kojouharov and Chen-Carpentier, 2004) to fit experimental data.

TRANSPORT MODELING

Supposing an homogeneous flow, the governing equation for solute transport in a saturated porous medium can be written as:

$$\frac{\partial c}{\partial t} + \mathbf{V} \cdot \nabla c - \nabla \cdot \left(\frac{\mathbf{D}}{\theta} \cdot \nabla c \right) = R(c) / \theta \, . \tag{11}$$

This equation clearly describes advective, dispersive and reactive terms. The advective term, depending on flow velocity, refers to the movement of the solute along with the flowing solvent in response to the hydraulic gradient. The dispersive term takes into account both, solute mixing due to local variation of transport velocity, and diffusion at molecular level. The reactive term would describe possible processes like adsorption, decay and reaction of the substances with other components. In what follows, we will assume that the process of adsorption onto the solid matrix is stationary, so only the fraction present in the liquid phase can move.

For a given porous medium, the characteristic advective time may be considered as

$$t_A = \delta / V \ ,$$

where δ is the correlation or characteristic length of the problem; this time is associated with the spreading of solute. The dispersive time is

$$t_d = \delta^2 / D \ ,$$

where D is the dispersion coefficient (a representative value of the dispersion tensor); this time is associated with mixing. Their physical interpretation is that after a time t_A the solute would be advected over a distance δ and after a time t_d it would be dispersed over the same distance. For the specific problems we are considering, δ corresponds to the mean radius of the packing material. The characteristic reaction time t_R depends on the reaction process involved and for our specific problem it will be defined later. The *Peclet number*

$$Pe \equiv t_d / t_A$$

gives us an idea of which processes, dispersion or advection, is the dominant one. $Pe >> 1$, implies that dispersion takes a long time for mixing in such a way that advection rules the transport of solute. The advective or dispersive *Damköhler numbers*

$$Da_A \equiv t_A / t_R$$

and

$$Da_d \equiv t_d / t_R \ ,$$

respectively, indicates whether the advection or dispersion is the dominant process compared with reaction. If $Da_A >> 1$ ($Da_d >> 1$) the reaction is faster than the advection (dispersion).

The method used to solve equation (11) depends on which is the dominant term. If reaction time t_R is much smaller than advective time t_A, the reaction is very fast, turning it difficult to simulate this process numerically because it would be very costly in computer time. To achieve this, Wheeler and Dawson (Wheeler and Dawson, 1987) proposed an *Operator Splitting* method. Each integration stage is done in two steps: the first solves the advection-diffusion equation without the reactive term, dealt with in the next step. For the advection-diffusion step, a finite difference or finite element scheme is used. For the second step, a standard integration method for ordinary differential equations as Runge-Kutta is usually applied.

Although in some cases it is possible to obtain analytical solutions for one dimensional flow for first order reactions (Cokca, 2003), in a more general situation numeric integration techniques are required to model the transport equation. The mathematical behavior of equation (11) changes from parabolic for dispersion dominated to hyperbolic for advection dominated problems. When advection governs the process, numerical instabilities like oscillations or numerical dispersion appear when the equation is discretized, giving rise to non-physical solutions. In these cases, to decompose the solution of the equation into two steps is also an option (Kojouharov and Chen-Carpentier, 2004; Kojouharov and Welfert, 2004). The most difficult part to solve is the advective - reactive step, which requires a method developed for this purpose. The dispersive step can be solved using a standard method.

As we mentioned above, the models of solute transport in porous media are described by means of concentrations averaged over many pore spaces, considering the porous media as continuous. But in flow processes with multiple reactive species, it is important to consider that the reaction is taking place at molecular level, where the velocity field is never homogeneous and the continuum hypothesis has no longer application.

Reactive mixing of dissolved compounds in porous media has recently received much attention (Kapoor et al., 1997; Kapoor et al., 1998, Raje and Kapoor, 2000; Gramling et al., 2002; Jose and Cirpka, 2004). In the laboratory, breakthrough experiments with columns packed with different materials simulating the porous media were performed for reactions of the type A + B $\rightarrow$ C. Some of these experiments use columns saturated with a solution of reactant B. Meanwhile, reactant A is injected, producing a displacement of the initial solution (Figure 1). In the contact area between both reactants, a third substance C is produced as a consequence of the reaction. Concentrations of reactants and products can be measured directly in the outflow of the column (Raje and Kapoor, 2000). Gramling et al. (2002) worked with a colorimetric instantaneous reaction and a solid matrix, adequately chosen to register dispersed light all along the column with a camera. Concentration data were obtained by image analysis. In another case, a set of probes was distributed along the column in order to measure concentrations using fiber – optics fluorometry (Jose and Cirpka, 2004).

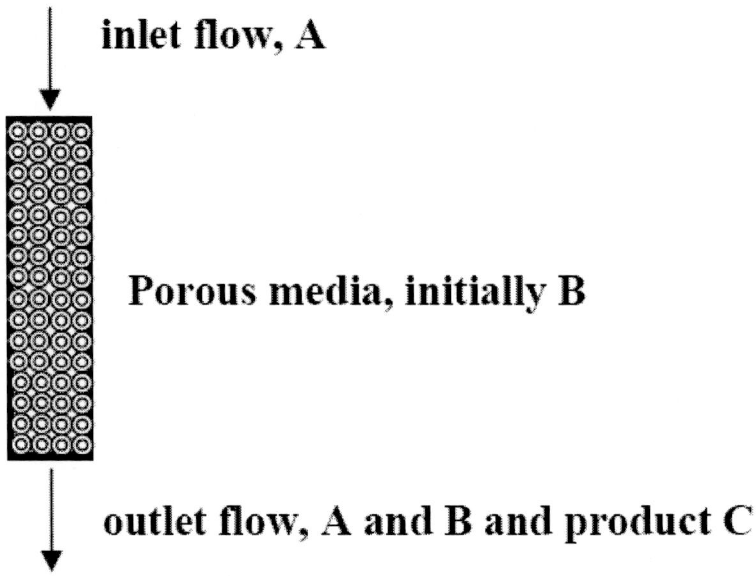

Figure 1. Initially reactant B fills the column, packed with a solid matrix. Reactant A is injected into the column, displacing the resident one and forming product C.

When dispersion time is much smaller than reaction time ($Da_d << 1$), the reactive substances achieve an homogeneous distribution before reacting; in this case, as expected, there are no significant differences between the experimental data and the numerical simulations based on the continuous hypothesis. This is not the case with the inverse time relation. As the flow details at poral level (microscale) are not known, estimated average concentrations (macroscale) are usually used. For each reactant, poral scale variabilities cause fluctuations around the volumetric average concentration, leading to segregation phenomena (Kapoor et al., 1997). As the distribution of the reactive species is not as homogeneous as required for a continuous hypothesis, the volume averaged mathematical continuum model seems to be limited to describe this chemical process, at least for short time scales. The time scales limit or introduce difficulties in the numerical simulations of the experiments because on the one hand, the significance of the segregation effect depends on them, and, on the other, these scales determine the dominant term and so, what the mathematical behavior of the equation will be like.

Bimolecular Reactive Transport

For one-dimensional bimolecular reactive transport flow through an isotropic and homogeneous medium, transport modeling with equation (11) leads to second order partial derivatives of a non-linear differential equation. Although the dynamic of the whole process depends on which is the prevailing time scale (i.e. advection, dispersion or reaction time) in equation (11), by one hand it is possible to develop a simple algorithm to fit the experimental data (Rubio et al., 2008) and by the other a consistent model of the segregation factor can be given.

For a reaction of the type A + B → C, the concentrations of reactants A and B are c_1 and c_2, respectively, and the concentration of the product C is c_3.

The source term for this reaction is $R = -\theta\,\Gamma c_1 c_2$, leading to:

$$\frac{\partial c_i}{\partial t} + V\frac{\partial c_i}{\partial x} - \frac{\partial}{\partial x}\left(D^*\frac{\partial c_i}{\partial x}\right) = -\Gamma c_1 c_2, \quad (i = 1, 2) \tag{12}$$

where $D^* = D/\theta$, the effective diffusion-dispersion coefficient, is assumed to be the same for all species. To study bimolecular reactive transport processes, a system of coupled equations, one for each reactant, must be solved. There is also an equation for the product in which the source term is $+\Gamma c_1 c_2$. Although this last equation is redundant, we include it in order to calculate directly c_3.

For a bimolecular reaction, the reaction time is defined as the depletion of one of the reactants (reactant A, for example): $t_R = (\Gamma c_2)^{-1}$, as a function of the concentration of the other. Although c_2 changes during the reaction, a first estimation of the minimum value of this time is $t_R = (\Gamma c_2^0)^{-1}$, where c_2^0 is the initial concentration of reactant B.

Reaction rate constants used in reactive transport modeling are usually determined in batch experiments where the reactants are well homogenized. These constants constitute a macroscopic parameter of the model. But the advection processes produce a differential movement at microscale that cannot be modeled under the continuum hypothesis. The lack of homogeneity at microscale makes the reactions less efficient, yielding to a lower product concentration than predicted by the continuous model. Due to unknown flow details or numerical limitations, usually mean concentrations (in macroscale) instead of punctual ones are known: $\overline{c}_i = c_i - c_i'$, where c_i' is the fluctuation respect of the mean value $\overline{c}_i$ of the *i* species. Variability at microscale induces a negative correlation of fluctuations known as segregation. It can be proved that including segregation implies considering an effective rate constant that can be modeled in the form $\Gamma_{eff} = \Gamma(1+s)$ (Kapoor et al., 1997), where the segregation factor *s* can only have values between -1 and 0. Generally, fluidistic models take the product of mean concentrations $R^* = -\Gamma\overline{c}_1\,\overline{c}_2$ instead of the mean product of concentrations:

$$R^* = -\Gamma\cdot\overline{c_1\cdot c_2} = -\Gamma\left(1+\left(\frac{\overline{c_1'\cdot c_2'}}{\overline{c_1}\cdot\overline{c_2}}\right)\right)\cdot\overline{c_1}\cdot\overline{c_2}\,. \tag{13}$$

Hence the reaction rate results larger than experimental measurements. At the start of the experiment the two reactants are highly segregated, so the segregation factor would be large and negative, but tends slowly to zero after enough time has elapsed, given that the dispersive dynamic of the flow homogenizes the reactants (Kapoor et al., 1997, Kapoor et al. 1998).

The experiments reported in the literature dealing with segregation were performed under very different conditions with respect to the characteristic times of the transport process. The one performed by Raje and Kapoor (2000) is governed by advection ($Da_A << 1$) and was solved by numerical simulation. Meanwhile, Gramling et al. (2002) have studied an almost instantaneous reaction ($Da_A >> 1$), proposing an analytical solution. While Raje and Kapoor's experiment can be solved with standard numerical techniques (Press et al., 1992), numerical simulation of the Gramling's experiment is difficult because the integration step must be smaller than t_R. In the last case reaction time step is several orders of magnitude smaller than the time step needed to integrate the advective process, requiring too much computing time. In both cases segregation is not included in the modeling, leading consistently to higher results than the experimental data.

A recent article (Meile and Tuncay, 2006) focuses on the study of scale-dependent reactive processes. The influence of the concentration variability at small (poral or micro) scales on reaction rates and the validity of the continuum approximation in reactive transport through a computer simulated porous medium were analyzed using a numerical model. They analyze different reactive processes –homogeneous, bimolecular and heterogeneous reactions. They also propose a correction term involving macroscopic concentration gradients and introducing a phenomenological parameter to improve the accuracy of reaction rates descriptions. Although this work is limited to diffusion driven transport, that is, the process is considered stationary ($\frac{\partial c}{\partial t} = 0$) and there is no advection ($V = 0$), their analysis seems to be a powerful approach to assess the problem of segregation in reactive transport.

Jose and Cirpka (2004) performed conservative and reactive experiments using fluorescent substances, in a homogeneously packed sand column. By means of fiber − optic fluorometry they were able to measure solute concentrations within the porous medium. Analyzing normalized conservative breakthrough curves (BTC) by the method of temporal moments, they derived mixing related dispersion coefficients. Predictions based on conservative breakthrough curves agreed well with measurements of reactive breakthrough curves. After comparison of BTCs they conclude that, in field scale applications, incomplete mixing on the poral scale would be of minor significance, but weak sorption might influence mixing significantly.

In a recent article published by Knutson et al. (2007), in a comparison between continuum and poral scale models, the authors suggest that better results could be obtained adjusting transversal dispersion coefficients (D_{trans}) instead of rate reaction (Γ). Nevertheless, the system is different from the examples considered here.

NUMERICAL SIMULATIONS

We have solved numerically a system of one-dimensional equations that models the bimolecular reactive − transport process corresponding to the experiments conducted by Raje and Kapoor (2000) and by Gramling *et al.* (2002) by means of a finite difference scheme, centered in space and forward in time that guarantees a precision of first order in time and second order in space (Press et al., 1992). Considering a uniform discretization in time $\{t_1, t_2, ..., t_N\}$, and space $\{x_1, x_2, ..., x_M\}$, the discretized equations can be written as:

$$
\begin{aligned}
c_{1i+1,j} &= c_{1i,j} - \frac{V\Delta t}{2\Delta x}\left(c_{1i,j+1} - c_{1i,j-1}\right) + \frac{D\Delta t}{\Delta x^2}\left(c_{1i,j+1} - 2c_{1i,j} + c_{1i,j-1}\right) - R_{i,j}, \\
c_{2i+1,j} &= c_{2i,j} - \frac{V\Delta t}{2\Delta x}\left(c_{2i,j+1} - c_{2i,j-1}\right) + \frac{D\Delta t}{\Delta x^2}\left(c_{2i,j+1} - 2c_{2i,j} + c_{2i,j-1}\right) - R_{i,j}, \\
c_{3i+1,j} &= c_{3i,j} - \frac{V\Delta t}{2\Delta x}\left(c_{3i,j+1} - c_{3i,j-1}\right) + \frac{D\Delta t}{\Delta x^2}\left(c_{3i,j+1} - 2c_{3i,j} + c_{3i,j-1}\right) + R_{i,j}.
\end{aligned} \tag{14}
$$

where Δt is the time step and Δx denotes step in space. The first subindex in $c_{1i,j}$ identifies the reactant, the other two denote the time instant and the position, respectively. For example, $c_{1i,j}$ is the concentration value for the reactant 1 at time t_i and position x_j, or, $c_{1i,j} = c_1(t_i, x_j)$. In what follows we will work with a macro-scale approach, so we are using c instead of $\overline{c}$, in order to simplify notations. The specific form of the reaction term $R_{i,j}$ will be discussed in the next section.

The stability of the solutions must be taken into account when solving numerically a mathematical problem (Gershenfeld, 1999). To achieve this for advective - diffusive - reactive problems, Morton and Mayers (1994) gave the following conditions:

$$
\left(\frac{V \cdot \Delta t}{\Delta x}\right)^2 < 2 \cdot \frac{D \cdot \Delta t}{\Delta x^2} < 1, \tag{15}
$$

where stability is related to the length of time and spatial integration steps; in our numerical simulations we take special care to verify this issue.

The main numerical difference between the two experimental data sets is the relationship between the characteristic times of advection and reaction. The problem analized by Raje and Kapoor is governed by advection ($t_A << t_R$). Meanwhile, Gramling et al. have studied an almost instantaneous reaction ($t_R << t_A$). A useful tool to avoid artificial instabilities that can be used with systems where reaction time is much smaller than advective time is provided by an *Operator Splitting* method (Press et al., 1992). There are different approaches for an operator splitting. One of them was proposed by Wheeler and Dawson (1987). The development of reaction and advection processes in time was splitted in two successive integration time steps. In the first step only advection (no reaction) is considered. The solution obtained is then used to integrate the equations with reaction (no advection) using a different time step.

We proposed a numerical scheme based on this idea (Rubio et al., 2008). The integration time step Δt was divided in two terms $\Delta t = \Delta t_1 + \Delta t_2$. First, equation (12) with no source was solved:

$$\frac{\partial c_i}{\partial t} + V\frac{\partial c_i}{\partial x} - D^* \frac{\partial^2 c_i}{\partial x^2} = 0, \tag{16}$$

considering a time step Δt_1. Then, with a time step Δt_2, the reaction term was solved considering neither transport nor dispersion, i.e.:

$$\frac{dc_i}{dt} = -\Gamma c_1 c_2, \tag{17}$$

where the solution of equation (16) is taken as the initial data. Given that in this case $t_R << t_A$, we could solve the equation (17) using a shorter integration time step Δt_2. From a physical point of view, it is necessary a Δt_1 time to mix solutes c_1 and c_2, then both react during a shorter time Δt_2 giving product c_3; after that, a new period Δt_1 is needed to mix and homogenize the reactants, and so on.

Because of the characteristics of the problem, appropriate time steps must be chosen to solve iteratively equations (16) and (17). An arbitrary integration step Δt_2 may yield wrong numerical results that would not fit the experimental data. A relationship between time steps that depends in a simple way on the characteristic times ($\Delta t_2 = \Delta t_1 / Da_A$) has been proposed (Rubio et al., 2008). Note that the ratio between the two intermediate time steps is given by the Damköhler number, indicating the relative weight of each term in the whole processes. The equations (16) and (17) must be repeatedly integrated one after the other until the final time is achieved.

To completely describe the process, boundary conditions must be given. At one end (top of the column), we assume that the solute c_1 is entering, hence neither the reactant c_2 nor the product c_3 are present. Thus, we imposed a Dirichlet type condition:

$$c_i(x,t)\big|_{x=0} = 0, \qquad i = 2, 3. \tag{18}$$

On the other hand, for the reactant c_1 we consider the condition

$$\left(Vc_1(x,t) - D\frac{\partial c_1(x,t)}{\partial x} \right)\Bigg|_{x=0} = Vc_1^0, \tag{19}$$

since c_1 enters the column at a fixed rate.

In order to use the same boundary conditions in both examples, the computational domain was taken long enough to be far apart from the bottom of the column. Then a Neumann condition

$$\left.\frac{\partial c_i(x,t)}{\partial x}\right|_{x\to\infty} = 0, \qquad i = 1, 2, 3, \tag{20}$$

was proposed for the three solutes, that is, no flux variation was assumed.

Segregation Intensity Model

Segregation intensity (s) modeling from first principles is not straightforward, so heuristic proposals (Meeder and Nieuwstadt, 2000; Meyle and Tuncay, 2006; Aguirre et al., 2006) are used when trying to reproduce experimental results. The reaction term is now an effective reaction term $\Gamma_{eff} = \Gamma(1+s)$.

Specifically, the function proposed (Meyle and Tuncay, 2006) is:

$$s = \frac{\overline{c_1' \cdot c_2'}}{\overline{c_1} \cdot \overline{c_2}} \approx \frac{\alpha}{\theta} \frac{\nabla \overline{c}_1 \cdot \nabla \overline{c}_2}{\overline{c}_1 . \overline{c}_2}, \tag{21}$$

where the parameter α depends on pore geometry, dispersivity and flux velocity, among other factors. In a different context, a similar proposal was used by Meeder and Nieuwstadt (2000) to describe dispersion from a point source in the atmospheric boundary layer; see also Aguirre *et al.*, (2006). Segregation intensity, representing the negative correlation between reactant concentrations, can be seen as the product of the reactant concentration gradient: when the concentration gradient is low, fluctuations will be small and segregation negligible; higher gradients will have a much larger effect on fluctuations, producing a stronger segregation effect.

In order to model a segregation factor for the reactive term, the parameter α may be estimated by minimizing the square errors:

$$J(\alpha) = \int (d(u) - c(u))^2 \, du\,, \tag{22}$$

where d is the data, c is the numerical approximation to the product concentration and u is the variable of integration. Note that in the experiments described by Raje and Kapoor the concentration of the product was measured at the end of the column as a function of time, while in Gramling's *et al.*, the measurements were taken along the column at different times. Hence, in the first case c and d do not depend on the position and the variable of integration is the time t. On the other hand, in the second case c and d do not depend on the time and the integration is performed along the column, thus the variable of integration is x.

The discretized reaction term may be written as:

$$R_{i,j} = \Gamma\left[1+\frac{\alpha}{\theta}\frac{\left(c_{1i,j+1}-c_{1i,j-1}\right)\left(c_{2i,j+1}-c_{2i,j-1}\right)}{\left(2\Delta x\right)^2 c_{1i,j}c_{2i,j}}\right]c_{1i,j}c_{2i,j}\Delta t\,. \tag{23}$$

As we mentioned before, for Raje and Kapoor's experiment, the concentration of the product was measured at the end of the tube, that is, at a fixed position x_{N+1}. Hence, the problem of minimizing (22) is now formulated as the minimization of the discretized functional

$$J(\alpha) = \sum_{i=1}^{t}\left(c_{3i,N+1}(\alpha)-d_i\right)^2, \tag{24}$$

where d_i is the experimental measured value of the product concentration at the time t_i and position x_{N+1}, and $c_{3i,N+1}(\alpha)$ is the simulated value for that concentration at the same time and position, for a specific parameter value α.

For Gramling's experiment, a fixed time t_k was considered, we seek for the value of $\overline{\alpha}$ that minimizes the discretized functional:

$$J(\alpha) = \sum_{j=1}^{m}\left(c_{3k,j}(\alpha)-d_j\right)^2, \tag{25}$$

where d_i is the experimental measured value of the product concentration at the time t_k and position x_j, and $c_{3k,j}(\alpha)$ is the simulated value for that concentration at the same time and position, for a specific parameter value α.

A Matlab code was developed to solve the equations for the concentrations and the estimation of the parameter α.

APPLYING THE MODEL: EXAMPLES AND DISCUSSION

The application of this segregation model, equation (21), to reactive transport reactions was tested analyzing published data from Raje and Kapoor (2000) and Gramling *et al.* (2002).

The value of $\overline{\alpha}$ clearly depends on the data considered when minimizing either (24) or (25). Thus, several simulations were performed by choosing different sets of data in order to obtain the estimated value $\overline{\alpha}$ that generates the concentration values that fits best the experimental data.

The experiment developed by Gramling *et al.* is based on a thin column packed with cryolite sand, with a refractive index similar to that of water. When saturated with water, the system becomes translucent and has negligible distortions for image analysis (CCD camera) along the tube. The chamber was initially saturated with EDTA (c_2); the reactant $CuSO_4$ (c_1) was pumped at the top of the column. The camera captured timed series of images during the development of the experiment, taking into account that an intense blue product was obtained. Using this setup, several experiments with different flow rates and dispersivities, but the same initial reactant concentrations, were performed.

The experimental setup of Raje and Kapoor was a column packed with glass beads. In this case, experiments with different initial reactant concentrations (1,2-naphthoquinone-4-sulfonic acid (c_1) and aniline (c_2)) and velocities were reported. The product (1,2-naphthoquinone-4aminobenzene) and the reactants were measured spectrophotometrically at the exit of the column.

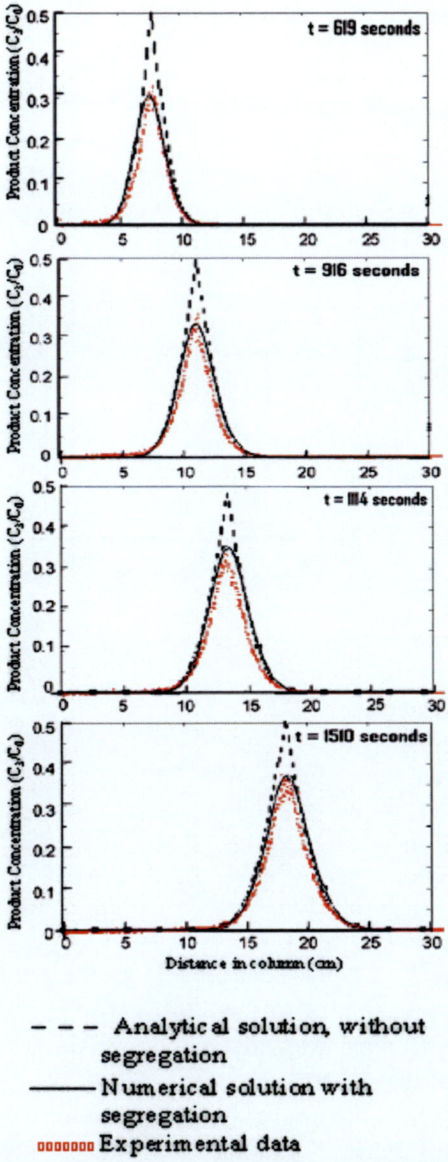

Figure 2. Product concentration along the tube at different times (Gramling *et al.* experiment). All the pictures correspond to the same experimental set up: flow velocity $V = 0.0121 cm/s$, dispersivity $\lambda = 0.145 cm$.

Figure 2 shows the results applying the previously described model to Gramling's experiment with $V = 0.0121\text{cm/s}$ and $\lambda = 0.145\text{cm}$, where the c_3 profile at four different times is shown. The solid line describes our numerical simulation, performed with $\alpha = 0.14$. Experimental data (points) are shown as well as the analytical solution without segregation proposed by Gramling (dashed line). Our simulation closely fitted the data (c_3 peak height and dispersion). The other relevant parameters of the experiments are: reaction rate constant $\Gamma = 2.3 \times 10^9 \text{M}^{-1}\text{s}^{-1}$, initial concentration $c_0 = 0.02\text{M}$ and porosity $\theta = 0.45$.

Figure 3 shows other Gramling's data for three experiments with different velocities and dispersivities, ordered according decreasing velocities. For the numerical simulations, the coefficient α used was of the same order of magnitude as the dispersivity ($\lambda = \frac{D}{V}$): $\alpha = 0.26$ (Fig. 3.a.), $\alpha = 0.17$ (Fig. 3.b.) and $\alpha = 0.14$ (Fig 3.c., see also Figure 2.c.). Although for these three experiments, a higher dispersivity value corresponds to an increased value of α, it is difficult to establish a clear correlation between these parameters because experiments were performed at different velocities and times.

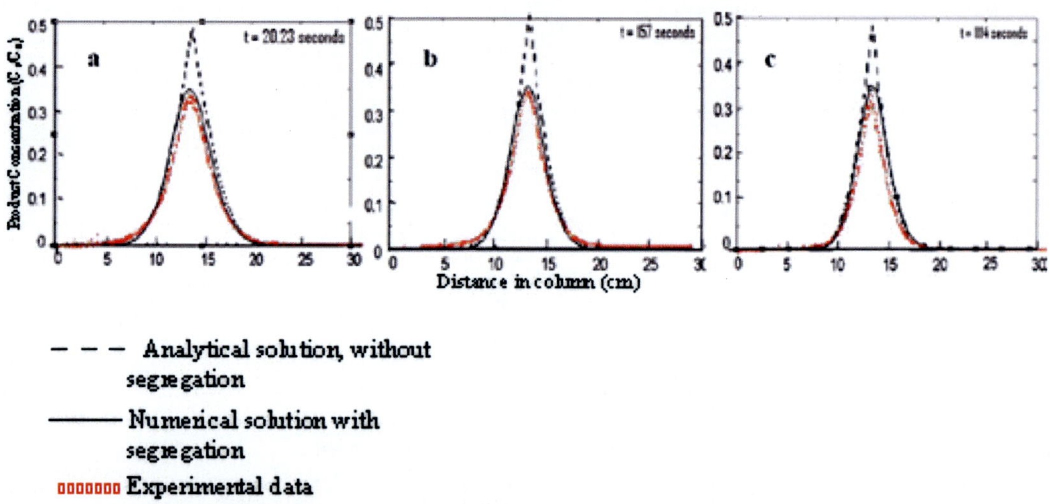

Figure 3. Product concentration along the tube at different times (Gramling *et al.* experiment). The pictures correspond to different experimental set ups: 3.a) $V = 0.67\text{cm/s}$, $\lambda = 0.261\text{cm}$. 3.b) $V = 0.0832\text{cm/s}$, $\lambda = 0.174\text{cm}$. 3.c) $V = 0.0121\text{cm/s}$, $\lambda = 0.145\text{cm}$.

A significative segregation can be expected for short times, as the dispersion and mixing of the reactants is incomplete. A priori, the product of the concentration gradients is who gives this information: at the beginning of the experience the concentration gradient of each reactant will be large, but with opposed signs (high negative correlation). On the other hand, Gramling *et al.* reported that the column was not repacked, but the dispersivity increase with increasing velocity. They suggest that probably this behaviour was linked to a decrease in the longitudinal spreading due to transversal molecular diffusion effects. At low velocities, Aris extension of Taylor's analysis (Bear, 1988) explained that dispersion decreases in tubes, suggesting that there is a nonlinear dependence between dispersion coefficient and velocity, Gramling *et al.* argued that a similar behaviour could be taking place in there experiments.

This mechanism could enhance mixing at low velocities in these experiments, leading to lower segregation and a smaller coefficient α, according to our results.

As the simulations shown in Figure 2 fitted well the experimental data obtained at different times, it is seems that there is not dependence between coefficient α and time.

Figure 4 shows the results corresponding to Raje and Kapoor data. The initial concentrations of reactants were: $c_1^0 = c_2^0 = 0.25$ mM and $c_1^0 = c_2^0 = 0.5$ mM, with seepage velocities of 0.07 m/s and 0.096 m/s, respectively. In these two cases, the dispersivity is the same and a correlation between α and V was observed ($\alpha = 1.4$ for $V = 0.07$ and $\alpha = 2$ for $V = 0.096$). This may indicate that when velocity is lower, reactants mix easier than with higher velocities, and segregation is lower. As concentration and velocity are the only investigated parameters in the experimental data, no further correlations can be obtained. The other relevant parameters of the experiments are: reaction rate constant $\Gamma = 438\mathrm{M}^{-1}\mathrm{s}^{-1}$ and porosity $\theta = 0.33$.

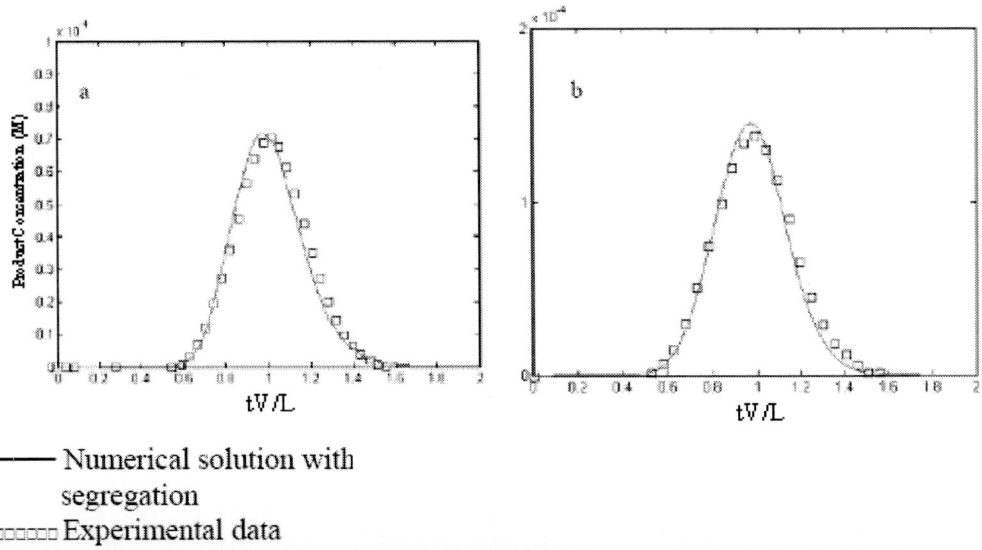

Figure 4. Product concentration at the bottom of the tube as a function of time (Raje and Kapoor experiment). The pictures correspond to different experimental set ups: 4.a) $\mathrm{V} = 0.07\mathrm{cm/s}$, $c^0 = 0.25\mathrm{mM}$. 4.b) $\mathrm{V} = 0.096\mathrm{cm/s}$, $c^0 = 0.5\mathrm{mM}$. Column length is $L = 18\mathrm{cm}$.

Comparing Gramling's and Raje and Kapoor's results, experimental parameters (λ, V, θ, R and also measurement times) are of the same order of magnitude in both cases. Nevertheless, we found differences of one order of magnitude for the coefficient α. Besides the uncertainties introduced by the minimization criteria used, that influence the coefficient α, the estimations are good as we can describe correctly the experimental data. Differences between both experiments are given mainly for the initial concentrations c_0 and rate constants Γ; although *a priori,* our model for s suggests that α does not depend on these parameters. Further studies would be needed in order to establish this fact.

Another parameter that takes very different values in both experimental setups is the molecular diffusion coefficient

$$D_{Gramling} \approx 10^{-7} \frac{\text{cm}^2}{s},$$

whereas

$$D_{Raje} \approx 10^{-5} \frac{\text{cm}^2}{s},$$

leading to a faster homogenization in Raje and Kapoor system. If this is so, the results are inconclusive, because we obtained higher segregation with these data, which is contradictory with a better homogenization. This result suggests that the diffusion coefficient would have a negligible influence on the coefficient α.

Here, experiments conducted under different conditions were analysed and simulated. The presented numerical results show a good performance of the model (21) for the segregation factor when the parameter α is estimated based on an experimental data. To achieve this end we minimize the square errors (22). The function (22) to be minimized takes the form (24) or (25) depending on the experiment under consideration. Although this model for the segregation is not new, it is the first time that an estimation parameter technique is used to obtain and estimator of the parameter model to improve the performance of the numerical solution of these types of problems. Moreover, the relationship between the best value $\overline{\alpha}$ of the estimator and the parameters of the experiments was discussed.

Further studies and more examples varying systematically the experimental conditions are needed for a better understanding of the segregation effects in advective – diffusive – reactive transport and the dependence of coefficient α on the macroscopic properties of the system. Nevertheless, the proposed method presents some benefits: it enables the simulation of experimental data, fitting well the results for different experimental set ups, and using flexible numerical techniques that allows to solve the transport equation with minimum computer requirements.

REFERENCES

Aguirre, A., Brizuela, A. B., Vonkovic, I. and Simoëns, S., 2006. A subgrid Lagrangian stochastic model for turbulent passive and reactive scalar dispersion. *International Journal of Heat and Fluid Flow*. 27, 627-635.

Ataie-Ashtiani, B., 2007. MODSharp: Regional-scale numerical model for quantifying groundwater flux and contaminant discharge into the coastal zone. *Environmental Modelling & Software*. 22, 1307-1315.

Ataie-Ashtiani, B. and Hosseini, S.A., 2005. Numerical errors of explicit finite difference approximation for two-dimensional solute transport equation with linear sorption. *Environmental Modelling & Software*. 20, 817-826.

Auger, L. and B. Legras; 2007 Chemical segregation by heterogeneous emissions, *Atmospheric Environment*. 41, 2303-2318.

Ayres, R., W. Schlesinger, R. Socolow, 1997, Human Impacts on the Carbon and Nitrogen Cycles; in: Industrial Ecology and Global Change, R. Socolow, C. Andrews, F. Berkhout and V. Thomas, Cambridge University Press.

Bear, J., 1988. Dynamics of fluids in Porous Media. Dover Publications, Inc. New York.

Brusseau, M.L., 1994, Transport of reactive contaminants in heterogeneous porous media, *Reviews of Geophysics*, 32, 3, 285-313.

Cao, J. and Kitanidis, P. K., 1998, Pore-Scale dilution of conservative solutes: An example. *Water Resour. Res.*, 34, 1941-1949.

Cokca, E., 2003. A computer program for the analysis of 1-D contaminant migration through a soil layer. *Environmental Modelling & Software*, 18, 147-153.

Dykaar, B. B. and Kitanidis, P. K., 1996, Macrotransport of biologically reacting solute through porous media. *Water Resour. Res.*, 32(2), 307-320.

Galmarini, S., Vilà-Guerau de Arellano and J.Duynkerke, P. G., 1995. The effect of micro-scale turbulence on the reaction rate in a chemically reactive plume. *Atmospheric Environment.* 29, 87-95.

Gelhar, L. W., 1993, *Stochastic Subsurface Hydrology*, Prentice-Hall, Englewood Cliffs, N. J.

Gershenfeld, N., 1999. The Nature of Mathematical Modeling. Cambridge University Press.

Gramling, C., Harvey, C. and Meigs, L., 2002. Reactive Transport in Porous Media: A Comparison of Model Prediction with Laboratory Visualization. *Environmental Science & Technology.* 36 (11), 2508-2514.

Hilst, G.R.; 1998, Segregation and chemical reaction rates in air quality models; *Atmospheric Environment.* 32(22), 3891-3895.

Huyakorn, P. S. and Pinder, G. F., 1983. Computational Methods in Subsurface Flow. Academic Press, Inc. (London) Ltd.

Jardine, P. M., S. E. Fendorf, M. A. Mayes, I. L.Larsen, S. C. Brooks and W. B. Bailey. 1999, Fate and Transport of Hexavalent Chromium in Undisturbed Heterogeneous Soil. *Environ. Sci. Technol.*, 33, 2939-2944.

Jørgensen, P.R., Urup, J., Helstrup, T., Jensen, M.B., Eiland, F., Vinther, F.P., 2004. Transport and reduction of nitrate in clayey till underneath forest and arable land. *Journal of Contaminant Hydrology*, 73, 207-226.

Jose, S. C. and Cirpka, O. A.; 2004, Measurement of Mixing Controlled Reactive Transport in Homogeneous Porous Media and Its Prediction from Conservative Tracer Test Data; *Environ. Sci. Technol.,* 38, 2089-2096.

Jury, W. A., Gardner, W. R. and Gardner, W. H., 1991. *Soil Physics*, John Wiley & Sons, Inc..

Kamra, S. K. and Lennartz, B., 2005. Quantitative indices to characterize the extent of preferential flow in soils. *Environmental Modelling & Software.* 20, 903-915.

Kapoor, V., Gelhar, L.W. and Miralles-Wilhelm, F., 1997, Bimolecular second-order reactions in spatially varying flows: Segregation induced scale-dependent transformation rates. *Water Res. Research.* 33 (4), 527-536.

Kapoor, V., Jafvert, C. T. and Lyn, D. A., 1998, Experimental Study of bimolecular Reaction in Pouseuille flow, *Water Res. Research.* 34 (8), 1997-2004.

Knutson, C., Valocchi, A. and Perth, C., 2007. Comparison of continuum and pore-scale models of nutrients biodegradation under transverse mixing conditions. *Advances in Water Resources.* 30, 1421-1431.

Köhne, J.M., Köhne, S., and Šimůnek, J., 2006. Multi-process herbicide transport in structured soil columns: experiments and model analysis. *Journal of Contaminant Hydrology.* 85, 1-32.

Kojouharov, H. V. and Chen-Charpentier, B. M., 2004. Nonstandard Eulerian-Lagrangian methods for multi-dimensional reactive transport problems. *Applied Numerical Mathematics.* 49, 225-243.

Kojouharov, H. V. and Welfert B. D., 2004. Generalized Nonstandard Numerical Methods for Nonlinear Advection-Diffusion-Reaction Equations, I. Lirkov et al. (Eds.): LSSC 2003, LNCS 2907, 465-472, Springer-Verlag Berlin Heidelberg

Laje Højberg, A., Engesgaard, P., and Bjerg, P.L., 2005. Pesticide transport in an aerobic aquifer with variable pH – Modeling of a field scale injection experiment. *Journal of Contaminant Hydrology.* 78, 231 – 255.

Logan, B., 1999. Environmental Transport Processes. John Wiley and Sons Inc.

Meeder, J. P. and Nieuwstadt, F. T. M., 2000. Large–eddy simulation of the turbulent dispersion of a reactive plume from a point source into a neutral atmospheric boundary layer. *Atmospheric Environment.* 34, 3563-3573.

Meile C. And Tuncay K., 2006. Scale dependence of reaction rates in porous media. Adv. in *Water Resources.* 29(1), 62-71.

Molemaker, M.J., and J. Vilà-Guerau de Arellano, 1998, Control of Chemical Reactions by Convective Turbulence in the Boundary Layer. *Journal of the Atmospheric Sciences.* 55, 568–579.

Morton, K.W. and Mayers, D.F., 1994. Numerical Solution of Partial Differential Equations. Cambridge University Press.

Press, H., Teukolsky, S., Vetterling, W. and Flannery, B., 1992. Numerical Recipies in Fortran 77: The art of scientific computing, (2nd edition). Cambridge Univ. Press;

Raje, D. S. and Kapoor, V., 2000, Experimental Study of bimolecular Reaction Kinetics in Porous Media. *Environ. Sci. Technol.* 34, 1234-1239.

Rashidi, M., Peurrung, L., Thompson, A. F. B. and Kulp, T. J., 1996, *Adv. Water Resour.* 19(3), 163-180.

Rubio, A.D., Zalts, A. and El Hasi, C.D., 2008*,* Numerical solution of the advection reaction diffusion equation at different scales. *Environmental Modelling & Software.* 23, 90-95.

Scheidegger, A.E., 1954. Statistical hydrodynamics in porous media. *J. Appl. Phys.* 25(8), 994–1001.

Steefel, C. I., DePaolo, D. J. and Lichtner, P. C., 2005, Reactive transport modeling: An essential tool and a new research approach for the Earth sciences. *Earth and Planetary Science Letters.* 240, 539-558.

Stein, A.F., V. Isakov, J. Godowitch, R. R. Draxler, 2007, A hybrid modeling approach to resolve pollutant concentrations in an urban area, *Atmospheric Environment.* 41, 9410-9426.

Stutter, M.I., Deeks, L.K., and Billett, M.F., 2005. Transport of conservative and reactive tracers through a naturally structured upland podzol field lysimeter. *Journal of Hydrology.* 300, 1-19.

Taylor, G. I., 1953. The dispersion of soluble matter flowing through a capillary tube. *Proc. Math. Soc. Lond.* 2:196-212.

Valocchi, A.J. and Malmstead, M., 1992. Accuracy of Operator Splitting for Advection-Dispersion-Reaction Problems. *Water Resources Research.* 28(5), 1471-1476.

Warneck, P.; 2000, Chemistry of the natural atmosphere, Academic Press, 2nd edition.

Wheeler, M. F. and Dawson, C. N., 1987. An operator-splitting method for advection-diffusion-reaction problems, Tech. Rep. 87-9, Dep. of Math. Sci., Rice Univ. Houston, Texas.

Wriedt, G. and Rode, M., 2006. Modeling nitrate transport and turnover in a lowland catchment system. *Journal of Hydrology*. 328, 157-176.

In: Porous Media: Heat and Mass Transfer... ISBN: 978-1-60692-437-2
Editors: J. L. Acosta and A.F. Camacho

Chapter 2

INDUCED POROELASTIC AND THERMOELASTIC STRESS CHANGES WITHIN RESERVOIRS DURING FLUID INJECTION AND PRODUCTION

Hamidreza Soltanzadeh and Christopher D. Hawkes
Department of Civil and Geological Engineering,
University of Saskatchewan, Canada

ABSTRACT

Stress changes induced by pore pressure or temperature changes occurring during fluid production (or injection) from (or into) porous reservoirs may lead to hazards such as ground movement, fluid leakage and earthquakes. Published data compiled for several reservoirs indicate that horizontal stress changes between 0.24 and 0.84 times the magnitude of the pore pressure change have been measured. Consequently, there is a clear need for models that predict pore pressure-induced stress changes. Temperature-induced stress changes, although less well documented, also have the potential to have significant effects. A review of three semi-analytical solutions for thermo-poroelasticity field equations is conducted in this chapter, covering: (1) the theory of strain nuclei; (2) the theory of inclusions; and (3) the theory of inhomogeneities. Closed-form solutions – most of which have been derived by the authors – are presented for pore pressure and thermally-induced stress changes within various geometrical variations of an ellipsoidal reservoir in a full space, for conditions where the mechanical properties of the reservoir are either identical to or different from the surrounding rock. Results for these models are presented in terms of dimensionless parameters, which facilitates their application to a broad range of reservoir conditions. A case study for the Ekofisk oil field, in the Norwegian sector of the North Sea, is used to demonstrate the use of these models. The results of this case study demonstrate that poroelastic modeling can predict horizontal stress changes with reasonable accuracy, and further challenge the common assumption that vertical stress changes are generally negligible.

1. Introduction

As a result of pore pressure or temperature changes, induced stress changes during fluid production from hydrocarbon reservoirs - or fluid injection for enhanced oil recovery or greenhouse gas sequestration - may lead to rock deformation, induced fracturing and fault reactivation. Potential consequences of these processes are ground movement and subsidence, well failures, reservoir fluid leakage and earthquakes. Due to the risks associated with these consequences, performance assessments of injection/production projects must include modeling of induced stress changes. This chapter focuses on the development and presentation of some relatively simple, yet powerful induced stress models, after first reviewing field data which demonstrate the occurrence and magnitudes of these stress changes.

2. Stress Change Measurement

Due to the difficulty of directly measuring vertical and maximum horizontal stresses in sedimentary basins, studies on stress change have mostly concentrated on measuring and predicting minimum horizontal stress ($\Delta\sigma_{Hmin}$) change - the magnitude of which can be measured more readily. The sensitivity of σ_{Hmin} to pore pressure change has long been recognized in the petroleum industry, most notably by people working on hydraulic fracturing (Santarelli et al., 1996). As such, there are several published case histories showing the relationship between pore pressure change and the change in minimum horizontal stress (see table 1).

To relate the minimum horizontal stress change to pore pressure change, a linear trend has been interpreted by many different investigators (e.g., Teufel & Rhett, 1991; Zoback & Zinke, 2002; Hawkes et al., 2005; Goulty, 2003). Such a relationship can be written as:

$$\Delta\sigma_{H\min} = K\Delta P \qquad (1)$$

A number of published values of K interpreted for different fields is presented in table 1. These results indicate that K ranges from 0.24 to 0.84. Table 1 shows a clear distinction between chalk and sandstone reservoirs. Chalk reservoirs typically have larger K-values than sandstone reservoirs. One might intuitively expect that, for a given lithology, K-values would increase with porosity. The data reported in table 1 seem to be generally consistent with this trend, although there are some discrepancies which cannot be resolved with the information available.

All values of K in table 1 were measured during fluid production operations (i.e., decreasing pore pressure). Due to the hysteretic character of reservoir rocks, a substantial difference between production and injection (i.e., loading and unloading of rock) effects is expected (Holt et al., 2004). A measurement of injection-induced stress change was reported by Santarelli et al. (1996) for an unnamed, poorly consolidated sandstone reservoir in the Norwegian sector of the North Sea, operated by Saga Petroleum. The results showed that no significant change in stress occurred in the reservoir (less than 4%) after injection. This result contrasted significantly with the K values of 0.42 and 0.70 interpreted for this reservoir during

Table 1. Pressure-induced stress change measurements within reservoirs

Field	Measured value of K	Porosity (%)	Reservoir Rock	Stress regime	Comments
Vicksburg Formation, south Texas	0.53 [(1), (2)] 0.48 [(3)]	21 [(4)]	Stacked deltaic sands[(1)]	Listric normal (growth) faults [(1)]	Stress-pressure depletion data taken from three fields, the McAllen Ranch, the Javellina and the McCook fields [(1)]
Ekofisk, centeral North Sea	0.8 on all three different regions: crest, flank, and outer flank [(5)] 0.84 ±0.02 on the crest and 0.82±0.09 on the flank and 0.77±0.06 on the outer flank [(6)]	30-48 [(7)]	High porosity chalk [(1)]	Normal [(7)]	The measurements were subdivided into three sections: crest, flank and outer flank of the reservoir [(7)]
Waskom, east Texas	0.46 [(1)]	7-16 [(8)]	Sandstone-rich sediments [(1)]	N/F (i.e., Not found by authors)	Data for the combined lithologies (sandstone, siltstone, limestone and shales) indicate a depletion response of 0.57 [(1)]
Magnus, U.K. sector of North Sea	0.68 [(1)]	21 [(9)]	Submarine fan sandstones in rotated fault blocks [(1)]	Normally faulted, present-day stress regime is uncertain [(1)]	
Valhall, Central North Sea	0.76 [(3)] 0.7 on the crest, 0.88 on the flank [(7)] 0.7 ±0.04 on the crest and 0.84±0.04 on the flank [(6)]	20-50 [(7)]	Soft chalk [(7)]	Normal [(7)]	
Unnamed in North Sea operated by Saga Petroleum	0.70 in the centeral part and 0.42 in the southern part [(10)]	N/F	Poorly consolidated sandstone [(10)]	N/F	$K \approx 0.04$ as interpreted during fluid injection.[(10)]
Gulf of Mexico – Field X	0.54 [(11)]	18-33 [(11)]	Deltaic sandstone [(11)]	Normal [(11)]	
McAllen Ranch field, Texas	0.5 [(2)]	14 [(12)]	Sandstone [(2)]	Normal [(1)]	
Venture Field, Nova Scotia	0.56 [(13)]	32	Sandstone [(14)]	N/F	

Table 1. (Continued).

Field	Measured value of K	Porosity (%)	Reservoir Rock	Stress regime	Comments
Rullison Field, Colorado	0.24 [3]	8 [15]	Sandstone [3]	N/F	Gas deposit is an aggregate of hundreds of separate small accumulations in discontinuous sandstone bodies. [16]
Rotliegend field, North Sea	0.69 [3]	15-20 [16]	Sandstone [3]	N/F	
Eldfisk, North Sea	0.84 [3]	17-50 [17]	Soft chalk [17]	N/F	
Wytch Farm Field, England	0.65 [1] , 0.55 [13]	10-18 [18]	Sandstone [1]	Normal [1]	

Legend:

N/F denotes "not found"

(I) Addis, 1997 (2) Khan & Tuefel, 1996 (3) Khan et al., 2000 (4) Al-Shaieb et al., 2000 (5) Tuefel & Rhett, 1991 (6) Goulty, 2003 (7) Zoback & Zinke, 2002 (8) Desbrandes & Yildiz, 1991 (9) Strachan et al., 2004 (10) Santarelli et al., 1996 (11) Chan & Zoback, 2002 (12) Tucker, 1979 (13) Wu & Addis, 1998 (14) Lubomir & Urrea, 1990 (15) Coffer et al., 1970 (16) LaFleur & Johnson, 1973 (17) Cook & Berkke, 2004 (18) Brown et al., 2000 .

production. The magnitude of this hysteresis effect may be considerably less in other reservoirs, depending for example on the degree of over- or under-consolidation and cementation of the reservoir rock, initial reservoir pressure (e.g., over-pressuring), magnitude of the pressure change, and reservoir depth.

STRESS ARCHING EFFECTS

3.1. Introduction

The assumption of uniaxial deformation has been one of the most popular approaches to model the poro-mechanical behaviour of reservoirs. For this type of model, the reservoir is considered to be constrained laterally, and to deform solely in the vertical direction. This model type is, strictly-speaking, only appropriate for laterally infinite reservoirs, but in a practical sense it is useful for reservoirs that are very thin relative to their lateral dimensions. Uniaxial models are not realistic, however, for reservoirs with aspect ratios (i.e., thickness/length ratios) that are not negligible, and/or if there is a significant contrast between the reservoir and the surrounding rock.

In effect, if a reservoir was a free body, effective stress changes would simply result in its contraction or expansion. However, the reservoir is "attached" to the surrounding rock, which works against the reservoir's tendency to contract or expand. Due to this competition between internal driving forces and external constraints, anisotropic changes in total stress may be induced, depending on reservoir geometry, mechanical property contrasts between the reservoir and surrounding rock, and the distribution of pore pressure within the reservoir. This phenomenon has been called arching (e.g., Mulders, 2003).

Due to the fact that there can be changes in vertical stress as well as horizontal stresses within and outside of a reservoir, arching ratios have been defined to give a more general explanation of stress change patterns. Arching ratios are most appropriate for thermo-poroelastic materials, where there is a linear relationship between the change in stress and the change in pore pressure or in temperature.

3.2. Poroelastic Arching Ratios

The ratio of the change in total stress to the effective change in pore pressure ($\alpha\Delta P$) within the reservoir has been referred to as the normalized poroelastic stress arching ratio by Soltanzadeh and Hawkes (2008), and it is expressed as follows:

$$\gamma_{\alpha(H_1)} = \Delta\sigma_{H_1} /(\alpha\Delta P)\,;\; \gamma_{\alpha(H_2)} = \Delta\sigma_{H_2} /(\alpha\Delta P)\,;\; \gamma_{\alpha(V)} = \Delta\sigma_V /(\alpha\Delta P) \qquad (2)$$

where $\gamma_{\alpha(H_1)}$, $\gamma_{\alpha(H_2)}$, and $\gamma_{\alpha(V)}$, respectively, are poroelastic normalized horizontal and vertical stress arching ratios; $\Delta\sigma_{H_1}$, $\Delta\sigma_{H_2}$, and $\Delta\sigma_V$, respectively, are horizontal and vertical stress changes; α is Biot's coefficient, and ΔP is the reservoir's pore pressure change.

Assuming that the poroelastic stress arching ratios for a given reservoir have been determined, the effective stress change within the reservoir can be calculated as follows:

$$\Delta\sigma'_{H_1} = -(1-\gamma_{\alpha(H_1)})(\alpha\Delta P);$$
$$\Delta\sigma'_{H_2} = -(1-\gamma_{\alpha(H_2)})(\alpha\Delta P); \qquad (3)$$
$$\Delta\sigma'_{V} = -(1-\gamma_{\alpha(V)})(\alpha\Delta P)$$

The value of the poroelastic normalized vertical stress arching ratio has commonly been considered negligible for reservoirs of large lateral extents (i.e., reservoirs that deform uniaxially). Unfortunately, in practice, there is no field measurement to assess the importance or magnitude of this effect. Analytical and numerical models do show, however, that arching may have a very significant effect leading to a reduction of the contractile strains within a reservoir and the redistribution of in-situ stresses. For instance, Mulders (2003) stated that, for small reservoirs made up of weak rocks, about 50% of the vertical effective stress change may be arched away (i.e., $\gamma_{\alpha(V)} \cong 0.5$). Using three-dimensional modeling, Kenter et al. (1998) interpreted that there was likely 20% to 30% vertical arching (i.e., $\gamma_{\alpha(V)} \cong 0.2$ to 0.3) during depletion of the Shearwater gas reservoir in the northern North Sea.

3.3. Thermoelastic Arching Ratios

For the sake of solving the problem for thermoelasticity, horizontal and vertical thermoelastic stress arching ratios (i.e., respectively $\gamma_{T(H_1)}$, $\gamma_{T(H_2)}$, and $\gamma_{T(V)}$) are defined as following:

$$\gamma_{T(H_1)} = \Delta\sigma_{H_1}/(\eta\Delta T),\ \gamma_{T(H_2)} = \Delta\sigma_{H_2}/(\eta\Delta T), \gamma_{T(V)} = \Delta\sigma_{V}/(\eta\Delta T) \qquad (4)$$

where η is the linear coefficient of thermal expansion, and ΔT is the temperature change within the reservoir. Due to the similar, dilatational nature of pore pressure change and temperature change effects, the thermoelastic solution can be considered as an analogous case for the poroelastic solution. As will be shown later in this chapter, the thermoelastic stress arching ratios can be evaluated by transformation of the poroelastic stress arching ratios.

4. Induced Stress Change Modeling

4.1. Background

Model types for geomechanical analysis of reservoirs may be categorized as analytical models, semi-analytical models and numerical models. Analytical models have been developed by using basic concepts of uniaxial poroelasticity (e.g., Hawkes et al., 2005), reservoir normal compaction (e.g., Goulty, 2003), or frictional equilibrium mechanisms (e.g.,

Holt et al., 2004) and are restricted to analyzing stress changes within a reservoir of idealized geometry and simplified fluid flow conditions

Semi-analytical models essentially use analytical solutions that can only be evaluated using numerical integration procedures. They are often capable of assessing stress changes both within a reservoir and in the surrounding rocks. These model types are also based on simplified geometrical and fluid flow assumptions, and they are usually developed using the assumption of linear poroelastic material behaviour for the reservoir and the surrounding rocks (e.g., Segall, 1985). There are three main theories upon which semi-analytical models for reservoir geomechanical analysis are based: (1) the theory of strain nuclei (e.g., Geertsma, 1966; Segall, 1985); (2) the theory of inclusions (e.g., Segall and Fitzgerald, 1998; Soltanzadeh and Hawkes, 2008); and (3) the theory of inhomogeneities (e.g., Rudnicki 1999, Soltanzadeh et al., 2007). The rest of this chapter describes the application of these theories to the analysis of strain and stress change fields induced by pore pressure change and thermal changes within reservoirs. The topic of induced stress change within the surrounding rock is beyond the scope of this chapter. Readers interested in this topic are referred to published works such as Segall (1985), Segall (1992), Segall et al. (1994), Soltanzadeh and Hawkes (2007), and Soltanzadeh et al. (2008).

To analyze more complicated reservoirs, accounting for more realistic geometries and rock/fluid behaviour, the use of numerical models is required. One of the most important advantages of some numerical models is their ability to model discontinuities, such as faults and weak shear zones that exist in the field.

4.2. Elasticity Field Equations

The equilibrium equations for a continuous medium are:

$$\sigma_{ij,j} + b_i = 0 \tag{5}$$

where σ_{ij} is the stress tensor and b_i represents body forces. In these equations, and those that follow, Einstein's index notation has been used. More information on index notations, which are commonly used to simplify the presentation of equations involving vector or tensor fields, can be found in most continuum mechanics textbooks (e.g., Mase, 1970). For a poroelastic medium, the relationship between the stress tensor and strain tensor (ε_{ij}) can be written as follows:

$$\sigma_{ij} = C_{ijkl}\varepsilon_{kl} - \alpha P \delta_{ij} - \beta_{ij}\Delta T \tag{6}$$

where C_{ijkl} is the elastic stiffness tensor, α is Biot's coefficient, P is pore pressure, δ_{ij} is the Kronecker delta, β_{ij} is the thermoelastic modulus tensor, and ΔT is the temperature change. For small strains, the relation between displacements (u_i) and strains is:

$$\varepsilon_{ij} = (u_{i,j} + u_{j,i})/2 \tag{7}$$

Substituting equations (6) and (7) in equation (5) and using the symmetrical property of elastic stiffness tensor (i.e., $C_{ijkl}=C_{klij}=C_{ijlk}$) results in the poroelasticity displacement field equation, as follows:

$$C_{ijkl}u_{k,li} - \alpha P_{,i} - \beta_{ij}\Delta T_{,j} + b_i = 0 \tag{8}$$

This equation can be written as the following general form:

$$C_{ijkl}u_{k,li} + f_i = 0 \tag{9}$$

where f_i is considered to be an equivalent force which, based on superposition, is defined as a linear combination of any or all of the following: body force (b_i), pore pressure effect ($-\alpha P_{,i}$), and thermal effect ($-\beta\Delta T_{,i}$).

In the case of isotropic linear elasticity, the stiffness tensor C_{ijkl} and thermoelastic modulus tensor β_{ij} are expressed as:

$$C_{ijkl} = \frac{2\mu\nu}{1-2\nu}\delta_{ij}\delta_{kl} + \mu(\delta_{ik}\delta_{jl} + \delta_{il}\delta_{jk}) \tag{10}$$

$$\beta_{ij} = 3K\eta\delta_{ij} = \frac{2\mu(1+\nu)}{1-2\nu}\eta\delta_{ij} \tag{11}$$

where K, μ, and ν, respectively, are bulk modulus, shear modulus and Poisson's ratio. Using equations (10) and (11), equation (8) will reduce to the following form for an isotropic poroelastic material:

$$\mu\nabla^2 u_i + \frac{\mu}{1-2\nu}u_{j,i}u_{j,j} - \frac{2\mu(1+\nu)}{1-2\nu}\eta T_{,i} - \alpha P_{,i} + b_i = 0 \tag{12}$$

4.3. Theory of Strain Nuclei

The theory of strain nuclei was essentially developed to solve the field equations in an elastic medium for point loading conditions, also called singularities, such as point forces, concentrated moments, and centers of dilatation (or compression). However, by integration, this methodology was extended to solve problems of distributed loading conditions (Love, 1944).

The simplest form of such a solution is for a point force in an infinite medium, which is known as Kelvin's problem. Imagine a force with unit magnitude in any direction acting on any point $\mathbf{x}'$ within an infinite medium. Since f_i in equation (9) has a body force nature, to imply the effect of this unit point force in equation (10), as an equivalent, we might use the Dirac delta function. The Dirac delta function is defined as follows, for a point $\mathbf{x}$ in an arbitrary volume V:

$$\begin{cases} \delta(\mathbf{x}-\mathbf{x}')=0 & if\ \mathbf{x}\neq\mathbf{x}' \\ \int_V \delta(\mathbf{x}-\mathbf{x}')dV=1 & if\ \mathbf{x}=\mathbf{x}' \end{cases} \tag{13}$$

In this case, equation (9) can be written as:

$$C_{knim}G_{ij,mn}(\mathbf{x},\mathbf{x}')+\delta(\mathbf{x}-\mathbf{x}')\delta_{jk}=0 \tag{14}$$

$G_{ij}(\mathbf{x},\mathbf{x}')$, which are known as Green's functions, are defined in this equation as the magnitude of displacement of point $\mathbf{x}$ in the i-direction, when a unit body force in the j-direction is applied at point $\mathbf{x}'$ in an elastic medium.

If, instead of a unit point load, a point force of F_j affects on point $\mathbf{x}'$, the displacement at point $\mathbf{x}$ can be found as:

$$u_i = F_j G_{ij}(\mathbf{x},\mathbf{x}') \tag{15}$$

In addition, if a distributed load f_j is applied on the volume Ω, the following solution can be used to solve the problem:

$$u_i = \int_\Omega f_j G_{ij}(\mathbf{x},\mathbf{x}')d\mathbf{x}' \tag{16}$$

Returning to the context of thermo-poroelasticity, f_j in equation (9) might be a consequence of loading conditions such as pressure or thermal changes. Since all deformations in these cases are volumetric, they are known as dilatational (or compressional) problems. For instance, in the case of pore pressure change, equation (16) can be written as:

$$u_i = \int_\Omega -\alpha P_{,j} G_{ij}(\mathbf{x},\mathbf{x}')d\mathbf{x}' \tag{17}$$

Integration of this equation by parts, results in the following equation (Segall, 1992):

$$u_i = \alpha \int_\Omega \Delta P G_i^D(\mathbf{x},\mathbf{x}')d\mathbf{x}' \tag{18}$$

where $G_i^D(\mathbf{x},\mathbf{x}')$ are called influence functions for dilatation and they are functions of Green's functions, as follows:

$$G_i^D(\mathbf{x},\mathbf{x}') = G_{ij,j}(\mathbf{x},\mathbf{x}') \tag{19}$$

Knowing u_i, equations (6) and (7) can be used to find induced stresses in the medium. These result in the following equation:

$$\Delta\sigma_{ij}(\mathbf{x}) = -\alpha \oint_{\Omega} \Delta P G_{ij}^{S}(\mathbf{x},\mathbf{x}')d\mathbf{x}' - \alpha\Delta P\delta_{ij} \tag{20}$$

where $G_{ij}^{S}(\mathbf{x},\mathbf{x'})$, which are referred to as stress functions, are a function of Green's functions as follows:

$$G_{ij}^{S}(\mathbf{x},\mathbf{x}') = \mu(G_{ip,pj}(\mathbf{x},\mathbf{x}') + G_{jp,pi}(\mathbf{x},\mathbf{x}')) + \frac{2\mu\nu}{1-2\nu}G_{kp,pk}(\mathbf{x},\mathbf{x}')\delta_{ij} \tag{21}$$

Green's functions for a full space can be derived by solving equation (14) using partial differential equation solutions such as a Fourier transform. Using the process of superposition, differentiation, and integration, and starting from the solution of the elasticity equations for a single force in a full space, the solution for influence functions for several different strain nuclei in an infinite medium or a semi-infinite medium can be derived (e.g., Mindlin, 1936; Mindlin and Cheng, 1950). An extensive list of Green's functions and influence functions was given by Sermet (2003).

The theory of strain nuclei was used by Geertsma (1966) to find the subsidence of reservoirs where the pore pressure change within the reservoir was considered uniform over the entire reservoir. Du and Olson (2001) applied this model in a discrete way to study subsidence for different arrangements of production wells. It was also used by Segall (1985) to analyse the stress distribution and fault reactivation potential in surrounding rocks of a depleting reservoir in Coalinga, California. Axisymmetric solution of the model had been introduced by Segall (1992) and was successfully applied to Lacq gas reservoir to analyse and predict the subsidence (Segall et al., 1994).

4.4. Theory of Inclusions

The theory of inclusions is another solution method for elasticity problems, which views such problems from a slightly different perspective than the theory of strain nuclei. By definition, an inclusion Ω is an embedded region in a homogeneous isotropic elastic matrix that would undergo an arbitrary strain (i.e., eigenstrain $\varepsilon_{ij}^{*}(\mathbf{x})$) if it was unbounded. Eigenstrains can be considered as internal displacements caused by various mechanisms, including poroelastic, plastic, and thermoelastic changes. Eshelby (1957) showed that such a problem is equivalent to solving the problem of body force distribution. Therefore, the solution for such a problem can be developed in a similar approach to the theory of nuclei strain by using Green's functions. In a general form, such a solution can be expressed as (Mura, 1982, p. 33):

$$u_i(\mathbf{x}) = -\oint_{\Omega} C_{jlmn}\varepsilon_{mn}^{*}(\mathbf{x}')G_{ij,l}(\mathbf{x},\mathbf{x}')d\mathbf{x}' \tag{22}$$

$$\Delta\sigma_{ij}(\mathbf{x}) = -C_{ijkl}\left\{\oint_{\Omega} C_{pqmn}\varepsilon_{mn}^{*}(\mathbf{x}')G_{kp,ql}(\mathbf{x},\mathbf{x}')d\mathbf{x}' + \varepsilon_{kl}^{*}(\mathbf{x})\right\} \tag{23}$$

where $\varepsilon_{ij}^{*}(\mathbf{x})=0$ for $x \notin \Omega$.

A significant discovery of Eshelby (1957) pertains to the behaviour of inclusions with ellipsoidal geometries embedded in a full-space matrix (i.e., neglecting the effects of ground surface on induced stresses and strains). He showed that, within such inclusions, for a uniformly applied eigenstrain, the induced strain and stress change fields are distributed uniformly throughout the inclusion. In such cases, the following relation exists between induced strain (ε_{ij}) and stress change ($\Delta\sigma_{ij}$) fields and eigenstrain vector:

$$\varepsilon_{ij} = S_{ijkl}\varepsilon_{kl}^{*} \tag{24}$$

$$\Delta\sigma_{ij} = C_{ijkl}(\varepsilon_{kl} - \varepsilon_{kl}^{*}) \tag{25}$$

where S_{ijkl} is Eshelby's tensor. Various sub-classes of an ellipsoidal inclusion geometry are considered in this work, including: oblate spheroid, prolate spheroid, sphere, and penny-shaped. Further, by allowing one or two axes to extend to infinity, an ellipsoidal geometry can be used to consider an inclusion that is either an elliptic cylinder or a layer of infinite lateral extent. The dilatational components of Eshelby's tensors for these different geometries are listed in Appendix A.

The theory of inclusions can be applied to a wide variety of problems in elasticity for different types of eigenstrains. However, in reservoir engineering, we are interested in dilatational eigenstrains resulting from pore pressure or temperature changes; the values of dilatational eigenstrain (ε^{C}) for these conditions, respectively, are:

$$\varepsilon_{C}^{*} = \frac{1}{3}\left(\frac{\alpha\Delta P}{K}\right) = \frac{(1-2\nu)}{2\mu(1+\nu)}(\alpha\Delta P) \tag{26}$$

and

$$\varepsilon_{C}^{*} = \eta\Delta T \tag{27}$$

Segall and Fitzgerald (1998) proposed using the theory of inclusions to find stress changes within and adjacent to an axisymmetric ellipsoidal shape in a full-space. Soltanzadeh and Hawkes (2008) applied this theory to find induced stress changes within and surrounding a reservoir of an arbitrary shape in a half-space medium.

Following, the theory of inclusions is applied to derive closed-form solutions for stress arching ratios within reservoirs with different variations of ellipsoidal geometries in a full-space. Soltanzadeh and Hawkes (2008) showed that, in a plane strain condition, a full-space solution might be used with a reasonable approximation if the reservoir depth and lateral extent are similar in magnitude, especially for reservoirs that are relatively thin (i.e., with an aspect ratio less than 0.2) which is the case for many geological reservoirs. Further, regardless of reservoir thickness, the full-space plane strain solution is virtually exact when reservoir depth is at least five times greater than lateral extent.

For an isotropic elastic medium, induced stress changes, hence poroelastic stress arching ratios, can be found by using equations (24), (25), and (26). Below, arching ratios are given for the above-noted classes of an ellipsoidal inclusion. A summary of these solutions is given in table 2.

For a prolate spheroid (i.e., an axisymmetric ellipsoid with a vertical dimension greater than the lateral dimension) with a vertical axis of symmetry, the poroelastic normalized stress arching ratios induced by pore pressure change (ΔP) can be determined as:

$$\begin{aligned} \gamma_{\alpha(H_1)} = \gamma_{\alpha(H_2)} &= \frac{1}{2}\frac{1-2\nu}{1-\nu}\left[1-\frac{1}{e^2-1}+\frac{e\cosh^{-1}e}{(e^2-1)^{3/2}}\right] \\ \gamma_{\alpha(V)} &= \frac{1-2\nu}{1-\nu}\left[\frac{e^2}{e^2-1}-\frac{e\cosh^{-1}e}{(e^2-1)^{3/2}}\right] \end{aligned} \tag{28}$$

where e is the aspect ratio of the prolate spheroid (the ratio of the vertical semi-axis to the horizontal semi-axis),which, by definition is always greater than one. [Note: Due to the fact that this axisymmetric solution has been derived for a full-space, by interchanging the axis indices, these equations could also be used for reservoirs with symmetry axes that are horizontal or inclined.]

For an oblate spheroid (i.e., an axisymmetric ellipsoid with aspect ratio (e) less than one) with a vertical axis of symmetry, the poroelastic normalized stress arching ratios (as shown previously by Fjær et al., 2008, p. 397) are calculated as follows:

$$\begin{aligned} \gamma_{\alpha(H_1)} = \gamma_{\alpha(H_2)} &= \frac{1}{2}\frac{1-2\nu}{1-\nu}\left[1+\frac{1}{1-e^2}-\frac{e\cos^{-1}e}{(1-e^2)^{3/2}}\right] \\ \gamma_{\alpha(V)} &= \frac{1-2\nu}{1-\nu}\left[\frac{e\cos^{-1}e}{(1-e^2)^{3/2}}-\frac{e^2}{1-e^2}\right] \end{aligned} \tag{29}$$

Poroelastic normalized stress arching ratios for a sphere (i.e., $e = 1$) can be found as:

$$\gamma_{\alpha(H_1)} = \gamma_{\alpha(H_2)} = \gamma_{\alpha(V)} = \frac{2}{3}\frac{1-2\nu}{1-\nu} \tag{30}$$

In the special case of an oblate spheroid when aspect ratio is very small ($e \le 0.2$), which is the case for many reservoirs, the inclusion shape can be approximated by a penny-shaped geometry with a maximum absolute error of 0.065. This maximum error occurs for the smallest possible value of Poisson's ratio; i.e., zero. For the penny-shaped case, poroelastic normalized stress arching ratios (as shown previously by Segall and Fitzgerald, 1998) can be derived as:

$$\gamma_{\alpha(H_1)} = \gamma_{\alpha(H_2)} = \frac{1-2\nu}{1-\nu}(1-\frac{\pi e}{4})\,;\; \gamma_{\alpha(V)} = \frac{1-2\nu}{1-\nu}\frac{\pi e}{2} \tag{31}$$

In the limiting case where aspect ratio (e) approaches zero, the ellipsoid resembles a horizontal layer with a very small thickness compared to its lateral extent. In such a case, the poroelastic normalized stress arching ratios are:

$$\gamma_{\alpha(H_1)} = \gamma_{\alpha(H_2)} = \frac{1-2\nu}{1-\nu}; \quad \gamma_{\alpha(V)} = 0 \tag{32}$$

These simple equations have been traditionally used for stress analysis in reservoir geomechanics. (e.g., Zoback and Zinke, 2002; Hawkes et al., 2005).

All the solutions given above have been developed for reservoirs with axisymmetric geometries. In cases where the reservoir has an elongated geometry an elliptic cylinder geometry can be used. This case gives a plane strain solution for the problem, as follows:

$$\gamma_{\alpha(H_1)} = \frac{1-2\nu}{1-\nu}\frac{1}{1+e}; \quad \gamma_{\alpha(H_2)} = \frac{1-2\nu}{1-\nu}; \quad \gamma_{\alpha(V)} = \frac{1-2\nu}{1-\nu}\frac{e}{1+e} \tag{33}$$

For the special case of a circular cylinder (i.e., e = 1) these equations reduce to:

$$\gamma_{\alpha(H_1)} = \gamma_{\alpha(V)} = \frac{1-2\nu}{2(1-\nu)}; \quad \gamma_{\alpha(H_2)} = \frac{1-2\nu}{1-\nu} \tag{34}$$

It is interesting to note that, for all of the geometries considered above, the following relationship between poroelastic normalized stress arching ratios holds true (Fjær et al., 2008, p. 398):

$$\gamma_{\alpha(H_1)} + \gamma_{\alpha(H_2)} + \gamma_{\alpha(V)} = \frac{2(1-2\nu)}{1-\nu} \tag{35}$$

Figure 1 shows the variation of horizontal and vertical normalized poroelastic stress arching ratios as a function of aspect ratio for different geometries, calculated using a value of 0.2 for Poisson's ratio (ν). Based on this figure, the maximum difference between results from a plane strain solution for an elliptic cylinder and an axisymmetric solution for a prolate spheroid is 0.13; this occurs when e = 1, where the geometries are a circular cylinder and a sphere, respectively. For aspect ratios less than 0.2, the maximum difference is 0.06. This shows that, for many reservoirs, either a plane strain or an axisymmetric solution might be considered without a significant error.

Dilatational eigenstrains in equations (26) and (27) and equations (24) and (25) can be used to find the relationship between thermoelastic and poroelastic stress arching ratios, as follows:

$$\gamma_{T(ij)} = \frac{2\mu(1+\nu)}{1-2\nu}\gamma_{\alpha(ij)} \tag{36}$$

Using the principle of superposition in elasticity, thermoelastic and poroelastic stress arching ratios can be applied in combination to find the induced stress changes in cases where both temperature and pore pressure changes have occurred.

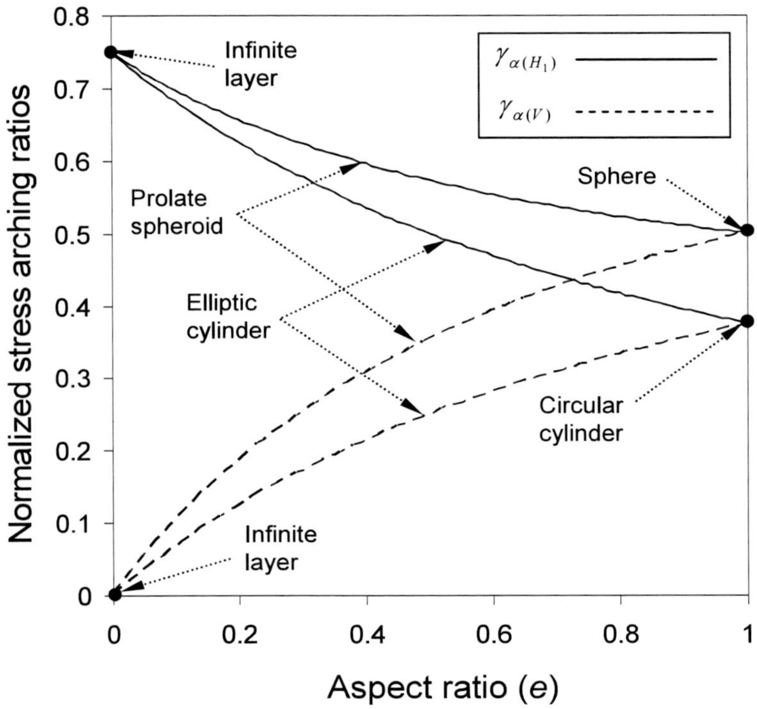

Figure 1. Poroelastic normalized horizontal ($\gamma_{\alpha(H_1)}$) and vertical ($\gamma_{\alpha(V)}$) stress arching ratios versus aspect ratio (e) for an elliptic cylinder (i.e., plane strain solution) and an oblate spheroid (i.e., an axisymmetric solution) for a Poisson's ratio of 0.2 when the material properties of the reservoir are identical to the surrounding rock. [Note: For the axisymmetric solution, $\gamma_{\alpha(H_1)} = \gamma_{\alpha(H_2)}$. For the plane strain solution, $\gamma_{\alpha(H_1)}$ represents the arching ratio in the cross-sectional plane, and $\gamma_{\alpha(H_2)}$ (the out-of-plane arching ratio) has a constant value of 0.75].

5. THEORY OF INHOMOGENEITIES

The mechanical property contrast between a reservoir and its surrounding rock may have a significant influence on the reservoir response. An extension of the theory of inclusions, referred to as the theory of inhomogeneities, can be used to deal with such cases (Rudnicki, 1999). The inclusion (i.e., reservoir) is called an inhomogeneity when it has different mechanical properties from the matrix (i.e., host rock) within which it is embedded. If eigenstrain ε^*_{ij} causes elastic strains of ε_{ij} in an inhomogneity, according to equation (25), the induced stress change within the inhomogeneity is:

$$\Delta\sigma_{ij} = C^*_{ijkl}(\varepsilon_{kl} - \varepsilon^*_{kl}) \tag{37}$$

where C^{*}_{ijkl} is the elastic stiffness matrix for the inhomogeneity. Eshelby (1957) proposed reducing the problem of an inhomogeneity for an ellipsoidal inhomogeneity by considering an equivalent inclusion which undergoes a fictitious eigenstrain ε^{**}_{ij}. In this case, the stress within the equivalent inclusion can be written as:

$$\Delta\sigma_{ij} = C_{ijkl}(\varepsilon_{kl} - \varepsilon^{**}_{kl}) \tag{38}$$

where C_{ijkl} is the elastic stiffness matrix for the matrix and the equivalent inclusion. The equivalency of both models requires the induced stress changes in equations (37) and (38) to be identical. This leads to the following equation to find the value of the fictitious eigenstrain:

$$\left[(C^{*}_{ijkl} - C_{ijkl})S_{klmn} + C_{klmn}\right]\varepsilon^{**}_{mn} = C^{*}_{ijkl}\varepsilon^{*}_{kl} \tag{39}$$

Upon finding values of the eigenstrain tensor from equation (39), equations (37) or (38) can be used to find the induced stress changes within the inhomogeneity.

Soltanzadeh et al. (2007) showed that, for the case of pore pressure change within a reservoir, equation (39) can be reduced to the following equation:

$$\left[\left(R_{\mu}C^{*}_{v(ijkl)} - C_{v(ijkl)}\right)S_{klmm} + C_{v(klmm)}\right]\varepsilon^{**}_{N(mm)} = \delta_{ij} \tag{40}$$

where R_{μ} is the ratio of shear moduli in the inhomogeneity and matrix (i.e., μ^{*}/μ), and $\varepsilon^{**}_{N(ij)}$ were referred to as the "normalized fictitious eigenstrains":

$$\varepsilon^{**}_{N(ij)} = \frac{\varepsilon^{**}_{ij}}{-(\alpha\Delta P/\mu)} \tag{41}$$

For an isotropic elastic material, C_{ijkl} and C^{*}_{ijkl} are defined as:

$$C_{ijkl} = \mu C_{v(ijkl)} \tag{42}$$

$$C^{*}_{ijkl} = \mu^{*} C^{*}_{v(ijkl)} \tag{43}$$

where:

$$C_{v(ijkl)} = \frac{2v}{1-2v}\delta_{ij}\delta_{kl} + \delta_{ik}\delta_{jl} + \delta_{il}\delta_{jk} \tag{44}$$

$$C^{*}_{v(ijkl)} = \frac{2v^{*}}{1-2v^{*}}\delta_{ij}\delta_{kl} + \delta_{ik}\delta_{jl} + \delta_{il}\delta_{jk} \tag{45}$$

and μ^{*} and v^{*} are, respectively, the shear modulus and Poisson's ratio of the inhomogeneity.

Knowing the values of $\varepsilon^{**}_{N(ij)}$, normalized poroelastic stress arching ratios within the inhomogeneity can be found as (Soltanzadeh et al., 2007):

$$\gamma_{\alpha(ij)} = C_{v(ijkl)}(\delta_{km}\delta_{lm} - S_{klmm})\varepsilon^{**}_{N(mm)} \tag{46}$$

Following, this theory is used to find stress arching ratios (hence induced stresses) within the various classes of ellipsoidal reservoirs. A summary of these solutions is given in table 2.

Soltanzadeh et al. (2007) used the theory of inhomogeneities to find a closed-form solution for a poroelastical elliptical reservoir with different materials from its surrounding rock, under plane strain conditions (i.e., an inhomogeneity that is an elliptic cylinder). The generalized form of their results can be written as follows:

$$\gamma_{\alpha(H_1)} = A_1 / A_4 \ ; \ \gamma_{\alpha(H_2)} = A_2 / A_4 \ ; \ \gamma_{\alpha(V)} = A_3 / A_4 \tag{47}$$

where $\gamma_{\alpha(H_1)}$ and $\gamma_{\alpha(H_2)}$ denote the in-plane and out-of-plane poroelastic normalized stress arching ratios, respectively, and:

$$A_1 = (1-2\nu^*)[R_\mu[e(1-2\nu)+2(1-\nu)]+e] \tag{48a}$$

$$A_2 = (1-2\nu^*)[R_\mu[R_\mu e(3-4\nu)+2(1+e^2)(1-\nu)]+e] \tag{48b}$$

$$A_3 = (1-2\nu^*)[R_\mu[2e(1-\nu)+1-2\nu]+1]e \tag{48c}$$

$$A_4 = R_\mu[2(1+e)^2(1-\nu)(1-\nu^*) - 2e\nu^*(1-2\nu) + R_\mu e(3-4\nu)] + e(1-2\nu^*) \tag{48d}$$

for the case of a circular cylinder (i.e., $e = 1$), these equations reduce to the following:

$$\gamma_{\alpha(H_1)} = \gamma_{\alpha(V)} = \frac{1-2\nu^*}{R_\mu + 1 - 2\nu^*} \ ; \ \gamma_{\alpha(H_2)} = \frac{(1+R_\mu)(1-2\nu^*)}{R_\mu + 1 - 2\nu^*} \tag{49}$$

In the case of a sphere, the results are (as shown previously by Rudnicki, 1999)

$$\gamma_{\alpha(H_1)} = \gamma_{\alpha(H_2)} = \gamma_{\alpha(V)} = \frac{2(1-2\nu^*)}{R_\mu(1+\nu^*) + 2(1-2\nu^*)} \tag{50}$$

and for an infinite layer:

$$\gamma_{\alpha(H_1)} = \gamma_{\alpha(H_2)} = \frac{1-2\nu^*}{1-\nu^*} \ ; \ \gamma_{\alpha(V)} = 0 \tag{51}$$

Unfortunately, deriving such simple closed-form solutions is not easy for all different ellipsoidal geometries. However, it is possible to find solutions with some simplifications in modeling assumptions. Sensitivity analyses by Rudnicki (1999) and Soltanzadeh et al. (2007) have shown that the values of stress arching ratios within reservoirs have only a weak dependency on Poisson's ratio of the matrix. Therefore, an assumption of identical Poisson's ratios in the inhomogeneity and the matrix (i.e., $\nu = \nu^*$) can be used, without incurring a significant error. By considering this simplification, for any axisymmetric ellipsoidal inhomogeneity including an oblate or prolate geometry, the following closed-form solutions can be found for poroelastic normalized stress arching ratios:

$$\gamma_{\alpha(H_1)} = \gamma_{\alpha(H_2)} = B_1 / B_3 , \ \gamma_{\alpha(V)} = B_2 / B_3 \tag{52}$$

where:

$$B_1 = (1+\nu)(1 - X_2 + (1 - R_\mu)X_1) + R_\mu(X_3 + \nu X_4) \tag{53a}$$

$$B_2 = (1+\nu)[1-(R_\mu - 1)X_1 - X_2] + R_\mu[(1-\nu)X_4 + 2\nu X_3] \tag{53b}$$

$$B_3 = (1+\nu)[(R_\mu - 1)^2 X_1 + (R_\mu - 1)X_2 + 1] \tag{53c}$$

where X_1, X_2, and X_3 are auxiliary variables which are functions of Eshelby tensor, as follows:

$$X_1 = (S_{1111} + S_{1122})S_{3333} - 2S_{3311}S_{1133} \tag{54a}$$

$$X_2 = S_{1111} + S_{1122} + S_{3333} \tag{54b}$$

$$X_3 = S_{3333} - S_{1133} \tag{54c}$$

$$X_4 = S_{1111} + S_{1122} - 2S_{3311} \tag{54d}$$

The relationships in equations (51) to (54) reduce to a more simplified form for a penny-shaped reservoir:

$$\gamma_{\alpha(H_1)} = \gamma_{\alpha(H_2)} = C_1 / C_3 , \ \gamma_{\alpha(V)} = C_2 / C_3 \tag{55}$$

where:

$$C_1 = (1-2\nu)\{\pi^2 e^2 (1 - R_\mu)(1+\nu) - 8R_\mu(1-\nu) - 2\pi e[R_\mu(\nu - 2) + 1]\} \tag{56a}$$

$$C_2 = \pi e(1-2\nu)[\pi e(1 - R_\mu)(1+\nu) - 2R_\mu(1-2\nu) - 2] \tag{56b}$$

$$C_3 = (R_\mu - 1)\pi e\{(1-2\nu)[2-\pi e(1-R_\mu)(1+\nu)] - R_\mu(3-4\nu)(1+\nu)\} - 8R_\mu(1-\nu)^2 \quad (56c)$$

Figure 2 shows the variation of normalized poroelastic horizontal and vertical stress arching ratios with respect to aspect ratio and shear modulus ratio for a reservoir shaped like an elliptic cylinder, and possessing the same Poisson's ratio (0.2 in this case) as the surrounding rock. Figure 3 shows the same parameters for a reservoir with a prolate spheroid shape. These figures show that the vertical stress arching ratio increases monotonically with increasing aspect ratio, whereas the horizontal stress arching ratio decreases monotonically with increasing aspect ratio only if shear modulus ratio is close to or greater than 1. In addition, it shows that both vertical and horizontal stress arching ratios decrease with increasing shear modulus ratio. An interesting fact in figure 2 is the variation of the out-of-plane stress arching ratio ($\gamma_{\alpha(H_2)}$) with respect to aspect ratio. From this figure, for a very small aspect ratio, $\gamma_{\alpha(H_2)}$ tends to be close to the value from equation (51) for an infinite layer, while with increasing aspect ratio it reaches a constant value which can be find from the formulation for a circular cylinder in equation (49).

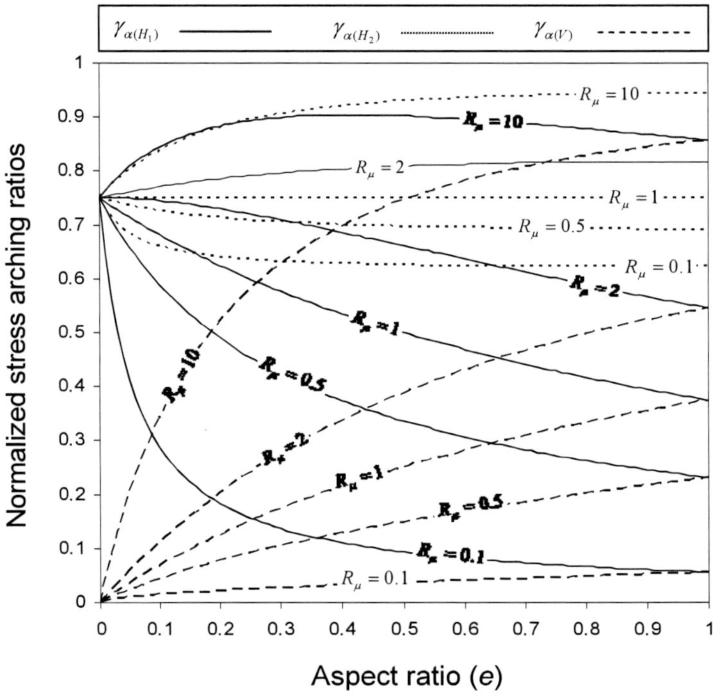

Figure 2. Normalized poroelastic in-plane horizontal ($\gamma_{\alpha(H_1)}$), out-of-plane horizontal ($\gamma_{\alpha(H_2)}$) and vertical ($\gamma_{\alpha(V)}$) stress arching ratios versus aspect ratio (e) for an elliptic cylinder (i.e., plane strain conditions). Poisson's ratios of the reservoir and the matrix are both 0.2 for this case.

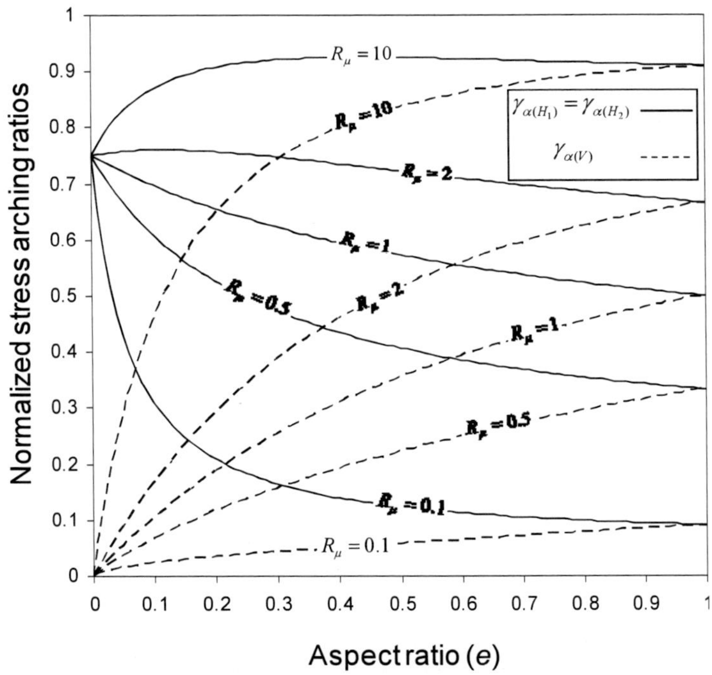

Figure 3. Normalized poroelastic horizontal ($\gamma_{\alpha(H_1)} = \gamma_{\alpha(H_2)}$) and vertical ($\gamma_{\alpha(V)}$) stress arching ratios versus aspect ratio (*e*) for an oblate spheroid (i.e., axisymmetric solution) for a Poisson's ratio ($v=v^*$) of 0.2 when the material properties of the reservoir is different from the surrounding rock.

Similar to the theory of inclusions, equation (36) can be applied to transform poroelastic stress arching ratios to their analogous thermoelastic arching ratios.

6. Case Study: Ekofisk Oil Field

Due to the relative wealth of published field data on the stress change magnitudes and indirect evidence of stress change during its producing life, the Ekofisk oil field is used here as a case study for which demonstrates the application of the solutions presented in this chapter. The Ekofisk Oil field is located in the Norwegian sector of the North Sea. It was discovered in 1969 and production began in 1971. In 1984, it was observed that the seabed over the Ekofisk Field had subsided by 3 m (Sulak, 1991). Although waterflooding commenced in this field in 1987, subsidence has continued at an almost constant rate (Rutledge et al., 1994) and it has experienced more than 10 m of subsidence (Dusseault et al., 2001). Also, there have been some casing failures in the shale caprock (Schwall & Denney, 1994) and appreciable gas leakage through the shale caprock, probably by flow through faults (Munns, 1985). All of this evidence suggests that stress change occurred due to pore pressure change within the reservoir.

Table 2. Poroelastic normalized stress arching ratios for different geometrical variations of an ellipsoidal reservoir derived using the theories of inclusions and inhomogeneities

Shape	Inclusion (Reservoir and surrounding rock have identical elastic properties)	Inhomogeneity (Elastic properties of reservoir are different from the surrounding rock)
Oblate spheroid 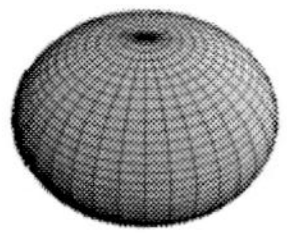	$\gamma_{\alpha(H_1)} = \gamma_{\alpha(H_2)} = \frac{1}{2}\frac{1-2\nu}{1-\nu}\left[1+\frac{1}{1-e^2}-\frac{e\cos^{-1}e}{(1-e^2)^{3/2}}\right]$ $\gamma_{\alpha(V)} = \frac{1-2\nu}{1-\nu}\left[\frac{e\cos^{-1}e}{(1-e^2)^{3/2}}-\frac{e^2}{1-e^2}\right]$	Numerical solution can be found from equations (40) to (46) - For a simplified inhomogeneity (i.e., $\nu=\nu^*$), a closed form solution can be found as: $\gamma_{\alpha(H_1)} = \gamma_{\alpha(H_2)} = B_1 / B_3$, $\gamma_{\alpha(V)} = B_2 / B_3$ $B_1 = (1+\nu)(1-X_2+(1-R_\mu)X_1)+R_\mu(X_3+\nu X_4)$ $B_2 = (1+\nu)[1-(R_\mu-1)X_1-X_2]+R_\mu[(1-\nu)X_4+2\nu X_3]$ $B_3 = (1+\nu)[(R_\mu-1)^2 X_1+(R_\mu-1)X_2+1]$ where: $X_1 = (S_{1111}+S_{1122})S_{3333}-2S_{3311}S_{1133}$ $X_2 = S_{1111}+S_{1122}+S_{3333}$; $X_3 = S_{3333}-S_{1133}$ $X_4 = S_{1111}+S_{1122}-2S_{3311}$
Prolate spheroid 	$\gamma_{\alpha(H_1)} = \gamma_{\alpha(H_2)} = \frac{1}{2}\frac{1-2\nu}{1-\nu}\left[1-\frac{1}{e^2-1}+\frac{e\cosh^{-1}e}{(e^2-1)^{3/2}}\right]$ $\gamma_{\alpha(V)} = \frac{1-2\nu}{1-\nu}\left[\frac{e^2}{e^2-1}-\frac{e\cosh^{-1}e}{(e^2-1)^{3/2}}\right]$	
Sphere 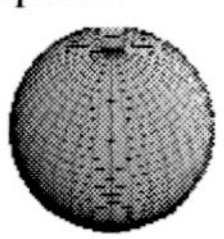	$\gamma_{\alpha(H_1)} = \gamma_{\alpha(H_2)} = \gamma_{\alpha(V)} = \frac{2}{3}\frac{1-2\nu}{1-\nu}$	$\gamma_{\alpha(H_1)} = \gamma_{\alpha(H_2)} = \gamma_{\alpha(V)} = \frac{2(1-2\nu^*)}{R_\mu(1+\nu^*)+2(1-2\nu^*)}$

Shape	Inclusion (Reservoir and surrounding rock have identical elastic properties)	Inhomogeneity (Elastic properties of reservoir are different from the surrounding rock)
Penny-shape 	$\gamma_{\alpha(H)} = \gamma_{\alpha(h)} = \frac{1-2\nu}{1-\nu}(1-\frac{\pi e}{4})$ $\gamma_{\alpha(V)} = \frac{1-2\nu}{1-\nu}\frac{\pi e}{2}$	Numerical solution can be found from equations (40) to (46) - For a simplified inhomogeneity (i.e., $\nu=\nu^*$), a closed form solution can be found as: $\gamma_{\alpha(H_1)} = \gamma_{\alpha(H_2)} = C_1 / C_3,\ \gamma_{\alpha(V)} = C_2 / C_3$ where: $C_1 = (1-2\nu)\{\pi^2 e^2(1-R_\mu)(1+\nu) - 8R_\mu(1-\nu) - 2\pi e[R_\mu(\nu-2)+1]\}$ $C_2 = \pi e(1-2\nu)[\pi e(1-R_\mu)(1+\nu) - 2R_\mu(1-2\nu) - 2]$ $C_3 = (m-1)\pi e$ $\{(1-2\nu)[2-\pi e(1-R_\mu)(1+\nu)] - R_\mu(3-4\nu)(1+\nu)\} - 8R_\mu(1-\nu)^2$
Elliptic cylinder 	$\gamma_{\alpha(H_1)} = \frac{1-2\nu}{1-\nu}\frac{1}{1+e}$ $\gamma_{\alpha(H_2)} = \frac{1-2\nu}{1-\nu}$ $\gamma_{\alpha(V)} = \frac{1-2\nu}{1-\nu}\frac{e}{1+e}$	$\gamma_{\alpha(H_1)} = A_1 / A_4, \gamma_{\alpha(H_2)} = A_2 / A_4, \gamma_{\alpha(V)} = A_3 / A_4$ $A_1 = (1-2\nu^*)[R_\mu[e(1-2\nu)+2(1-\nu)]+e]$ $A_2 = (1-2\nu^*)[R_\mu[R_\mu e(3-4\nu)+2(1+e^2)(1-\nu)]+e]$ $A_3 = (1-2\nu^*)[R_\mu[2e(1-\nu)+1-2\nu]+1]e$ $A_4 = R_\mu[2(1+e)^2(1-\nu)(1-\nu^*)$ $-2e\nu^*(1-2\nu)+R_\mu e(3-4\nu)]+e(1-2\nu^*)$
Circular cylinder 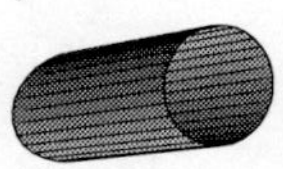	$\gamma_{\alpha(H_1)} = \gamma_{\alpha(V)} = \frac{1}{2}\frac{1-2\nu}{1-\nu}; \gamma_{\alpha(H_2)} = \frac{1-2\nu}{1-\nu}$	$\gamma_{\alpha(H_1)} = \gamma_{\alpha(V)} = \frac{1-2\nu^*}{R_\mu+1-2\nu^*}, \gamma_{\alpha(H_2)} = \frac{(1+R_\mu)(1-2\nu^*)}{R_\mu+1-2\nu^*}$
Infinite layer	$\gamma_{\alpha(H_1)} = \gamma_{\alpha(H_2)} = \frac{1-2\nu}{1-\nu}; \gamma_{\alpha(V)} = 0$	$\gamma_{\alpha(H_1)} = \gamma_{\alpha(H_2)} = \frac{1-2\nu^*}{1-\nu^*}\ ;\ \gamma_{\alpha(V)} = 0$

6.1. Reservoir Characteristics

The Ekofisk structure is an elongated anticlinal structure with a principal fold axis in an approximately north-south direction. The depth to the crest of the anticline is approximately 2840 m. The areal extent of the reservoir is 6.8×9.3 km (Lewis et al., 2003). The average thickness of the oil column is 300 m, including both the Ekofisk and Tor formations (Goulty, 2003). According to in-situ measurements, reservoir pressure has been depleted in the crestal area of the field from approximately 45 MPa in 1975 to 25 MPa in 1990. During this time, minimum horizontal stress decreased from approximately 51 MPa to 35 MPa. A linear relationship has been observed between minimum horizontal stress and reservoir pressure over this time period, with a K value of 0.8 (Teufel & Rhett, 1991). Stated in terms consistent with the material presented in this chapter, and using a value of 1 for Biot's coefficient (see below), this means that the horizontal normalized stress arching ratio for the Ekofisk reservoir is 0.8.

6.2. Geomechanical Properties

There is a broad scatter of values interpreted for the elastic properties of the Ekofisk reservoir and surrounding rocks reported in the literature. Proposed values for Young's modulus of the reservoir vary between 0.05 GPa to 2.2 GPa, where values for the surrounding rocks vary from 1.1 GPa to 14 GPa (Gutierrez & Hansteen, 1994; Boade et al., 1989; Lewis et al., 2003). Proposed values for Poisson's ratio vary from 0.15 to 0.25 for the reservoir and 0.20 to 0.42 for the surrounding rocks (Gutierrez & Hansteen, 1994; Boade et al., 1989; Lewis et al., 2003; Segall & Fitzgerald, 1998). As a rough estimation, based on the literature data and the results of back analyses conducted by the authors, it does not seem unreasonable to consider a contrast between reservoir rock and surrounding rocks which correspond to a shear modulus ratio (R_μ) equal to 0.1. This means that the shear modulus of the surrounding rocks is 10 times larger than the reservoir's shear modulus. Also, a value of 0.20 for the Poisson's ratio of the reservoir is deemed to be a reasonable estimate. Given the aforementioned lack of sensitivity of stress arching ratios to Poisson's ratio of the surrounding rock, a value of 0.20 was also assumed for this parameter. Based on laboratory experiments, Teufel & Rhett (1991) proposed a value of 1 for Biot's coefficient for the highly fractured Ekofisk reservoir rocks.

6.3. Induced Stress Change Analysis

To obtain a first estimate of induced stress change, the solution developed based on the theory of inclusions may be used (identical material properties for both the reservoir and the surrounding rock). Substituting a reservoir aspect ratio (e) of 0.044 (i.e., 300 m / 6800 m) for the Ekofisk field in equations (31) and (33), the poroelastic horizontal and vertical stress arching ratios are found to be, respetively, 0.73 and 0.05 for a penny-shape reservoir and 0.72 and 0.03 for a reservoir with an elliptic cylinder geometery. This estimated value of horizontal stress arching ratio is not very different from the measured value of 0.8.

Given the apparent contrast between material properties of the reservoir and surrounding rocks, it would be more reasonable to use the theory of inhomogeneities for the Ekofisk reservoir. Due to the small aspect ratio of this reservoir, the use of this full-space solution is deemed acceptable. Using a reservoir aspect ratio of 0.044 in equation (47) for a plane strain solution, the horizontal and vertical normalized stress arching ratios are found to be 0.80 and 0.18, respectively. Axisymmetric solutions given for a prolate spheroid or penny-shape reservoir gives the values of 0.82 and 0.26, respectively, for poroelastic horizontal and vertical stress arching ratios.

These values of the horizontal arching ratio (0.80 and 0.82) are very close to the value found by Teufel & Rhett (1991). However, Teufel & Rhett (1991) and other authors (e.g., Zoback & Zink, 2002; Goulty, 2003) implicitly assumed that there has been no change in vertical stress in the Ekofisk reservoir stress because of its large lateral extent relative to its thickness. This assumption differs markedly from the result obtained using the poroelastic models presented in this chapter; i.e., a predicted vertical stress change that is 18% or 26% of the pore pressure change. Clearly, field measurements of vertical stress changes would be beneficial for assessing the accuracy of these models' predictions, which may have significant implications in some reservoirs.

7. Conclusion

A summary of induced stress changes in several sandstone and chalk reservoirs has been given. Horizontal stress changes from 0.24 to 0.84 times the magnitude of the pore pressure change have been measured, with chalks typically showing larger stress changes. A review on semi-analytical models for predicting induced stress changes in a poroelastic medium, including the theory of strain nuclei, the theory of inclusions, and theory of inhomgeneites have been conducted.

Applying the theory of inclusions, semi-analytical solutions have been developed for induced stress changes in various classes of ellipsoidal reservoir geometries in a homogeneous, poroelastic full-space. Furthermore, Eshelby's theory of inhomogeneities has been used to develop closed-form solutions for these same geometries, accounting for material property contrasts between the reservoir and surrounding rocks. A summary of all of the solutions has been provided in the form of two reference tables. Results for both models have been presented in terms of dimensionless parameters, which facilitates their application to a broad range of reservoir dimensions and pressure-change magnitudes in reservoirs that conform reasonably well to the idealized geometries used for these models. A simple relationship between thermoelastic and poroelastic stress changes has been presented, which enables the calculation of thermally induced stress changes using the formulations presented in this chapter for normalized poroelastic stress arching ratios. A case study of the Ekofisk field has demonstrated the practical validity and feasibility of the presented models, as well as the need to also consider the effects of vertical stress changes for certain reservoir geometries and material properties.

8. NOMENCLATURE

b_i	Body force vector
C_{ijkl}	Elastic stiffness matrix
e	Reservoir aspect ratio (height/width)
G_{ij}	Green's functions
K	Bulk modulus
P	Pore pressure
R_μ	Shear modulus ratio (μ^*/μ)
S_{ijkl}	Eshelby's tensor
T	Temperature
u_i	Displacement field
α	Biot's coefficient
β_{ij}	Thermoelastic modulus tensor
$\gamma_{\alpha(ij)}$	Poroelastic normalized stress arching ratios
$\gamma_{T(ij)}$	Thermoelastic stress arching ratios
η	Linear coefficient of thermal expansion
ε_C^*	Dilatational eigenstrain
δ_{ij}	Kronecker delta
$\delta(\mathbf{x})$	Dirac delta function
ε_{ij}	Strain tensor
ε_{ij}^*	Eigenstrain tensor
μ	Shear modulus
ν	Poisson's ratio
σ'_{ij}	Effective stress tensor
σ_{ij}	Stress tensor

APPENDIX A. DILATATIONAL COMPONENTS OF ESHELBY'S TENSOR FOR DIFFERENT GEOMETRICAL VARIATIONS OF AN ELLIPSOIDAL INCLUSION (AFTER MURA, 1982)

Eshelby's Tensor	Spheroid	Sphere	Penny-shape	Elliptic cylinder
	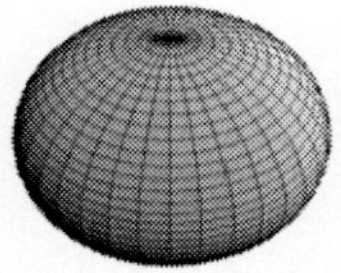 	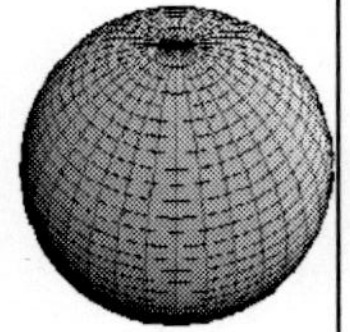		
S_{1111}	$\frac{3}{8(1-\nu)}\left[1-\frac{1+3F}{2(e^2-1)}\right]+\frac{1-2\nu}{4(1-\nu)}(1+F)$	$\frac{7-5\nu}{15(1-\nu)}$	$\frac{13-8\nu}{32(1-\nu)}\pi e$	$\frac{1}{2(1-\nu)}\left[\frac{e^2+2e}{(1+e)^2}+(1-2\nu)\frac{e}{1+e}\right]$
S_{1122}	$\frac{1}{8(1-\nu)}\left[1-\frac{1+3F}{2(e^2-1)}\right]-\frac{1-2\nu}{4(1-\nu)}(1+F)$	$\frac{5\nu-1}{15(1-\nu)}$	$\frac{8\nu-1}{32(1-\nu)}\pi e$	$\frac{1}{2(1-\nu)}\left[\frac{e^2}{(1+e)^2}-(1-2\nu)\frac{e}{1+e}\right]$
S_{1133}	$\frac{1}{4(1-\nu)}\frac{e^2(1+3M)}{e^2-1}-\frac{1-2\nu}{4(1-\nu)}(1+F)$	S_{1122}	$\frac{2\nu-1}{8(1-\nu)}\pi e$	$\frac{\nu}{1-\nu}\frac{e}{1+e}$
S_{2211}	S_{1122}	S_{1122}	S_{1122}	$\frac{1}{2(1-\nu)}\left[\frac{1}{(1+e)^2}-(1-2\nu)\frac{1}{1+e}\right]$
S_{2222}	S_{1111}	S_{1111}	S_{1111}	$\frac{1}{2(1-\nu)}\left[\frac{1+2e}{(1+e)^2}+(1-2\nu)\frac{1}{1+e}\right]$
S_{2233}	S_{1133}	S_{1122}	S_{1133}	$\frac{\nu}{1-\nu}\frac{1}{1+e}$

(Continued).

Eshelby's Tensor	Spheroid 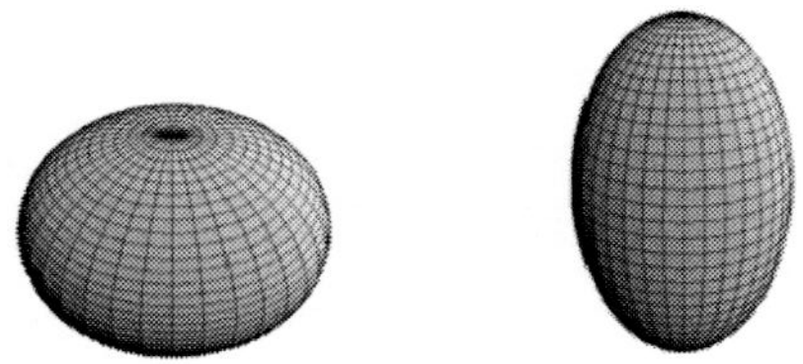	*Sphere* 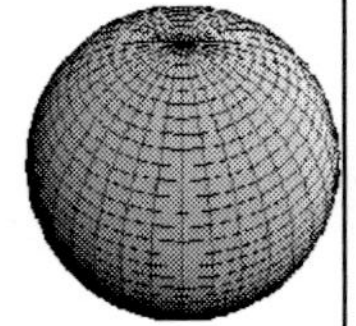	Penny-shape 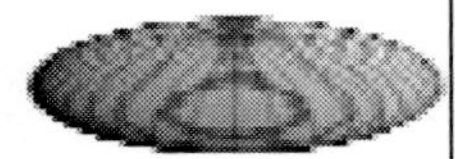	Elliptic cylinder 
S_{3311}	$\frac{1}{4(1-\nu)}\frac{1+3F}{e^2-1}+\frac{1-2\nu}{2(1-\nu)}F$	S_{1122}	$\frac{\nu}{1-\nu}\left(1-\frac{4\nu+1}{8\nu}\pi e\right)$	0
S_{3322}	S_{3311}	S_{1122}	S_{3311}	0
S_{3333}	$\frac{1}{2(1-\nu)}\left[1-\frac{e^2(1+3F)}{e^2-1}\right]-\frac{1-2\nu}{2(1-\nu)}F$	S_{1111}	$1-\frac{1-2\nu}{1-\nu)}\frac{\pi e}{4}$	0
Comments	For $e>1$ (i.e., a prolate spheroid): $F=1/(e^2-1)-(e\cosh^{-1}e)/(e^2-1)^{3/2}$ For $e<1$ (i.e., an oblate spheroid): $F=1/(e^2-1)+(e\cos^{-1}e)/(1-e^2)^{3/2}$	$e=1$	$e\leq 0.2$	Plane strain geometry

REFERENCES

Addis, M.A. The stress depletion response of reservoirs. *SPE Annual Technical Conference and Exhibition* 1997, 5-8 October, San Antonio, Texas. SPE 38720.

Al-Shaieb, Z.; Puckette, J.; Patchett, J.; Deyhim, P.; Li, H., Close A., Birkenfeld R. Identification and characterization of reservoirs and seals in the Vicksburg formation, TCB Field, Klebberg County, Texas. *As a report for the Gas Research Institute,* July 2000.

Boade, R.R.; Chin, L.Y.; Siemers, W.T. Forecasting of Ekofisk reservoir compaction and subsidence by numerical simulation. *Journal of Petroleum Technology* 1989, 41(7): 723-728. SPE 17855.

Brown, G.A.; Kennedy, B.; Meling, T. Using fibre-optic distributed temperature measurements to provide real-time reservoir surveillance data on Wytch Farm Field horizontal extended-reach wells, *SPE Annual Technical Conference and Exhibition,* 2000, 1-4 October. Dallas, Texas, SPE 62952.

Chan, A.W.; Zoback, M.D. Deformation Analysis in Reservoir Space (DARS): A simple formalism for prediction of reservoir deformation with depletion. *SPE/ISRM Rock Mechanics Conference* 2002, 20-23 October, Irving, Texas. SPE 78174.

Coffer, H.F.; Frank, G.W.; Bray, B.G. Project Rulison and the economic potential of nuclear gas stimulation. *Gas Industry Symposium* 1970, 21-22 May, Omaha, Nebraska. SPE 2876.

Cook, C.C.; Berkke, K. Productivity preservation through hydraulic propped fractures in the Eldfisk North Sea chalk field. *SPE Reservoir Evaluation & Engineering* 2004, 7(2): 105-114. SPE 88031.

Desbrandes, R.; Yildiz, T. Field applications of wireline formation testers in low-permeability gas reservoirs. *SPE Gas Technology Symposium* 1991, 22-24 January, Houston, Texas. SPE 21502.

Du, J.; Olson, J.E. A poroelastic reservoir model for predicting subsidence and mapping subsurface pressure fronts. *Journal of Petroleum Science and Engineering* 2001, 30:181–197.

Dusseault, M.B.; Bruno, M.S.; Barrera, J. Casing shear: causes, cases, cures. *SPE Drilling & Completion* 2001, 16(2):98-107.

Eshelby, J.D. The determination of the elastic field of an ellipsoidal inclusion and related problems, *Proceeding of the Royal Society of London* 1957, A241(1226): 326-396.

Fjær, E.; Holt, R.M.; Horsrud, P.; Raaen, A.M.; Risnes, R. *Petroleum related rock mechanics – 2nd edition*, Elsevier, Amsterdam, The Netherland. 2008. pp 395-398.

Geertsma, J. Problems of rock mechanics in petroleum production engineering. *Proceeding of 1st Congress of the International Society of Rock Mechanics* 1966, National Laboratory of Civil Engineering, Lisbon, Vol. I: 585–594.

Goulty, N.R. Reservoir stress path during depletion of Norwegian chalk oilfields. *Petroleum Geoscience* 2003, 9(3): 233-241.

Gutierrez M.; Hansteen, H. Fully coupled analysis of reservoir compaction and subsidence. *European Petroleum Conference* 1994, 25-27 October, London, U.K. SPE 28900.

Hawkes, C.D.; McLellan, P.J.; Bachu, S. Geomechanical factors affecting geological storage of CO2 in depleted oil and gas reservoirs. *Journal of Canadian Petroleum Technology* 2005, 44(10): 52-61.

Holt, R.M.; Flornes, O.; Li, L.; Fjær, E. Consequences of depletion-induced stress change on reservoir compaction and recovery. *The 6th North America Rock mechanics Symposium (NARMS): Rock Mechanics across Borders and Disciplines* 2004, June 5-9, Houston, Texas.

Kenter, C.J.; Blanton, T.L.; Schreppers G.M.A.; Baaijens, M.N.; Ramos, G.G. Compaction study for Shearwater Field. *SPE/ISRM Rock Mechanics in Petroleum Engineering* 1998, 8-10 July, Trondheim, Norway. SPE 47280.

Khan, M.; Tuefel, L.W. Prediction of production-induced changes in reservoir stress state using numerical model. *SPE Annual Technical Conference and Exhibition* 1996, 6-9 October. Denver, Colorado, SPE 36697.

Khan, M.; Teufel, L.W.; Zheng, Z.; Baker, A. Determining the effect of geological and geomechanical parameters on reservoir stress path through numerical simulation. *SPE Annual Technical Conference and Exhibition* 2000, 1-4 October, Dallas, Texas. SPE 63261.

LaFleur, K.K.; Johnson, A.K. Well stimulation in the North Sea: a survey. *SPE European Meeting* 1973, 2-3 April, London, UK. SPE 4315.

Lewis R.W.; Makurat, A.; Pao, W.K.S. Fully coupled modeling of seabed subsidence and reservoir compaction of North Sear oil fields. *Hydrogeology Journal* 2003, 11(1): 142-161.

Love, A.E.H. *A treatise on the mathematical theory of elasticity*; Dover: New York, NY. 1944. pp 186–189.

Lubomir, F.J.; Urrea, V.H.N. Geology and diagenetic history of overpressured sandstone reservoirs, Venture gas field, offshore Nova Scotia, Canada. *AAPG Bulletin* 1990, 74(10): 1640-1658.

Mase, G. E. *Continuum mechanics, Schaum's Outline Series.* McGraw-Hill: New York, NY. 1970.

Mindlin, R.D. Force at a point in the interior of a semi-infinity solid, *Physics* 1936, 13: 195-202.

Mindiin, R.D.; Cheng D.H. Nuclei of strain in the semi-infinite solid, *Journal of Applied Physics* 1950, 21: 926-930.

Mulders, F.M.M. *Modelling of stress development and fault slip in and around a producing gas reservoir.* Ph.D. Thesis, Delft University of Technology: Netherlands. 2003.

Munns, J.W. The Valhall Field: A geological overview. *Marine and Petroleum Geology* 1985, 2(1): 23-43.

Mura, T. *Micromechanics of defects in solids.* Martinus Nijhoff Publishers: The Hague, Netherlands, 1982.

Rudnicki, J.W. Alteration of regional stress by reservoirs and other inhomogeneities: stabilizing or destabilizing? *Proc. 9th Int. Congr. on Rock Mech* 1999, Paris, France, Swets & Zeitlinger: Lisse, Netherlands, 3:1629-1637.

Rutledge, J.T.; Fairbanks, T.D.; Albright, J.N.; Boade, R.R.; Dangerfield, J.; Landa, G.H. Reservoir microseismicity at the Ekofisk oil field. *Eurock 94: International Conference on Rock Mechanics in Petroleum Engineering* 1994, 29 August-1 September, Balkema, Rotterdam. SPE 28099.

Santarelli, F.J.; Tronvoll, J.T.; Svennekjaier, M.; Skeie, H.; Henriksen, R.; Bratli, R.K. Reservoir stress path: The depletion and the rebound. *SPE/ISRM Rock Mechanics in Petroleum Engineering* 1996, 8-10 July, Trondheim, Norway. SPE 47350.

Seremet, V.D. *Handbook of Green's Functions and Matrices*. WIT Press: Boston, 2003.

Schwall, G.H.; Denney, C.A. Subsidence induced casing deformation mechanisms in the Ekofisk Field. *Rock Mechanics in Petroleum Engineering* 1994, 29 August-1 September, Balkema, Rotterdam, Netherland. SPE 28091.

Segall, P. Stress and subsidence resulting from subsurface fluid withdrawal in the epicentral region of 1983 Coalinga earthquake. *Journal of Geophysical Research* 1985, 90: 6801-6816.

Segall, P. Induced stresses due to fluid extraction from axisymmetric reservoirs. *Pure & Applied Geophysics* 1992, 139: 535–560.

Segall, P.; Grasso, J.R.; Mossop, A. Poroelastic stressing and induced seismicity near the Lacq gas field, southwestern France. *Journal of Geophysical Research* 1994, 99(B8): 15423–15438.

Segall, P.; Fitzgerald, S.D. A note on induced stress change in hydrocarbon and geothermal reservoirs. *Tectonophysics* 1998, 289: 117-128.

Soltanzadeh, H.; Hawkes, C.D. Semi-analytical models for stress change and fault reactivation induced by reservoir production and injection. *Journal of Petroleum Science and Engineering* 2008, 60: 71-85.

Soltanzadeh, H.; Hawkes, C.D.; Sharma, J. Closed-form solutions for production- and injection-induced stresses in plane-strain reservoirs with elastic properties different from the surrounding rock, *International Journal of geomechanics*. 2007, 7(5): 353-361.

Strachan, C.J.; Heath, S.M.; White, K.; & Williams, G. Experience with pre-emptive squeeze treatments on BP Magnus with aqueous based scale inhibitors. *SPE International Symposium on Oilfield Scale* 2004, 26-27 May, Aberdeen, UK. SPE 87462.

Sulak, R.M. Ekofisk Field: the first 20 years. *Journal of Petroleum Technology* 1991, October: 1265-1271. SPE 20773.

Teufel, L.W.; Rhett, D.W. Geomechanical evidence for shear failure of chalk during production of the Ekofisk field. *66th Annual Technical Conference and Exhibition of the Society of Petroleum Engineering,* 1991, 6-9 October, Dallas, TX. SPE 22755.

Tucker, R.L. Practical pressure analysis in evaluation of proppant selection for the low-permeability highly geopressured reservoirs of the McAllen Ranch (Vicksburg) Field. *Symposium on Low Permeability Gas Reservoirs* 1979, 20-22 May, Denver, CO. SPE 7925.

Wu, B.; Addis, M.A. Stress estimation in faulted regions: The effect of residual friction. *SPE/ISRM Rock Mechanics in Petroleum Engineering* 1998, *8-10 July,* Trondheim, Norway, SPE 47210.

Zoback, M.D.; Zinke, J.C. Production-induced normal faulting in the Valhall and Ekofisk oil fields. *Pure & Applied Geophysics* 2002, 159: 403-420.

In: Porous Media: Heat and Mass Transfer...
Editors: J. L. Acosta and A.F. Camacho

ISBN: 978-1-60692-437-2

Chapter 3

POROUS HYDROGELS

M. Pradny, J. Michalek and J. Sirc

Institute of Macromolecular Chemistry, Academy of Sciences of the Czech Republic, Heyrovsky Sq. 2, 162 06 Prague 6, Czech Republic
Centre for Cell Therapy and Tissue Repair, Charles University, Prague, Czech Republic

ABSTRACT

Porous materials are inorganic or organic crosslinked or uncrosslinked contain pores of almost all sizes. The chapter is focused only on porous hydrophilic gels (hydrogels) with pore size from units to hundreds of micrometers, which are generally called macroporous or superporous. The macroporous hydrogels are most often used in the biomedical field, e.g. for tissue engineering, cell therapy and as biocompatible implants.

Porosity is usually achieved by one of the four methods:

Crosslinking polymerization in the presence of substances that are solvents for the monomers, but precipitants for the formed polymer.

Crosslinking polymerization in the presence of water soluble substances (sugars, salts) which are washed out from the hydrogel after polymerization.

Crosslinking polymerization in the presence of substances relasing gases which remain in the resulting hydrogel.

Freeze-sublimation of the hydrogel swollen in water.

By using any of the procedures, hydrogels with communicating or non-communicating pores can be obtained, i.e., with the pores that are or are not interconnected. This chapter is focused on the second-mentioned method, which makes it possible to adjust the pore size in a wide range and to prepare materials with communicating and non-communicating pores. Other methods allow control the pore size only in a small range; however, they are useful for some purposes as well. For most applications communicating pores are preferred.

Abbreviations

HEMA	2-hydroxyethyl methacrylate
MA	methacrylic acid
MaNa	sodium salt of methacrylic acid
MOETACl	[2-(methacryloyloxy)ethyl]trimethylammonium chloride
EOEMA	2-ethoxyethyl methacrylate
HPMA	*N*-(2-hydroxypropyl)methacrylamide
EDMA	ethylene di(methacrylate)

1. Introduction

Recently, porous hydrogels have been frequently mentioned in connection with tissue engineering, cell therapy, chromatographic columns, adsorbents and membranes. Figure 1 shows a number of works published on this topic since 1997; the last year reflects the end of February 2008. It seems that continuously increasing trend can be expected.

Evidently, most of the articles are focused on porous hydrogels as scaffolds for tissue engineering and cell therapy. The cell cultivation on them is usually successful due to their good biocomatibility and a high specific surface area. Moreover, biodegradable hydrogel scaffolds [1] are able to disappear after cell cultivation and, consequently, the newly created tissue is continuous.

Generally, porous materials are non-homogenous organic or inorganic substances of natural or. The chapter is focused just on a narrow group of porous materials based on hydrophilic crosslinked polymers prepared with communicating (interconnected) pores. The structure of communicating pores plays a key role in most of the applications in many fields.

2. Classification of the Porous Hydrogels by Pore Size

The porous materials can be classified from various points of view –by the origin (natural x synthetic), composition (organic x inorganic), mechanical or swelling properties, etc. The main property affecting consequent characteristics of porous materials is their physical structure, particularly size, shape and total volume of pores and walls.

The termsof porous materials, sometimes called "sponges", found in the literature are inconsistent. Only the nanoporous [2] (according to IUPAC) materials are consistently divided into synthetic origin (wood, rocks). Non-homogeneity of these materials consists in wall separated pores microporous (pore size < 2 nm), mesoporous (2<pore size<50 nm) and macroporous (50<pore size<1000 nm). However, very often materials containing pores in the range of tens and hundreds [3-7] or even more thousands [8] of micrometers are intentioned as macroporous as well. The materials containing pores larger than 100 μm are called superporous [9-27] and materials with the ratio of the pore and particle diameter higher than 0.01 are called gigaporous [28]. The walls are in this case composed of particles (see 3.1.)

The bunchs of nanofibers [29] or microfibers [30] are also porous materials, where the fibers can be considered as walls and surrounding spaces as pores. In contrast to "classical" porous hydrogels, where a part of pores is always non-communicating, fibrous structures are characteristed by communication of absolutelly all pores. An interesting example of nanoporous hydrogel is nanofibers of copolymer HEMA-EOEMA (figure 2) prepared by electrospinning in the presence of poly(ethylene oxide), which was subsequently washed out with water [31]. Similar nanofibers were prepared from cellulose triacetate [32]. This material is a promising superabsorbent due to its large specific surface area, even higher than nanofibers themselves.

3. Preparative Methods for Porous Hydrogels

3.1. Crosslinking Polymerization in the Presence of Substances that Are Solvents for Monomers, but Precipitants for Formed Polymer

The porous material prepared by crosslinking polymerization of monomers in the solvent that is a precipitant for the resulting polymer. The polymer precipitates usually in the form of sphere particles of nano- or micrometer dimensions [33, 34], which subsequently form the walls. The required distance between the discrete particles can be obtain by an appropriate concentration of monomer(s). The pores are then formed by the spaces between the particles or their aggregates. The main advantage of this technique is the formation of very thin walls required for superabsorbing materials [35]. The pores usually communicate and the fraction of non-communicating (disconnecting) pores is very low. This method of preparation is the simplest compared with the others and does not require any highly sophisticated apparatus. A disadvantage is just a limited adjustment of pore sizes, because any small changes in the polymerization mixture (necessary for controlling the resulting properties) affect the pore sizes. For example, we prepared such polymer by copolymerization of HPMA with a crosslinker (EDMA or *N,N′*-methylenebisacrylamide) in acetone/dimethyl sulfoxide as a porogen mixture [33]. A typical structure of such porous hydrogel is shown in figure 3.

3.2. Crosslinking Polymerization in Presence of Soluble Substances (Particles of Sugars, Salts) which Are Washed out from the Hydrogel after Polymerization

The water-soluble particles are added to a mixture of monomer, crosslinker and initiatior. After polymerization, the particles are washed out with water [1, 5, 19, 36 - 38]. The controlled size of these particles corresponds to the size of pores. An advantage of this technique is a simple adjustment of pore size and relatively easy characterization [5]. A limitation of this method is a higher thickness of pore walls. Using this method, prepare porous hydrogels with communicating (interconnecting) or noncommunicating (disconnecting) pores can be prepared. The percentage of communicating pores increases with increasing amount of soluble particles (porogen) in the polymerization mixture. This effect is followed by deterioration of mechanical properties. The morphology of materials prepared using this technique is obviously different from the morphology of materials prepared according to the 3.1. (figure 4). The copolymerization of HEMA-*co*-MOETACl with

EDMA as the crosslinker is an example of preparation of hydrogel by this method. The sodium chloride particles of average sizes 15, 40 and 70 μm were used as a porogen.

It was necessary to use a fourteen times higher amount of the porogen (sodium chloride) relative to the sum of the polymerization components to achieve communicating pores. For exclusion of bubbles and homogenization of monomers with sodium chloride oligo(ethylene glycol) (MW 300) was added. This component is a solvent for monomers but not for sodium chloride. After thorough mixing of all components, the viscous paste was polymerized in the apparatus depicted in figure 5. However, the preparation of porous hydrogels from the pre-prepared polymers is in principle possible but complicated and little reproducible [39]. The problem arises from sedimentation of salt particles in the polymerization mixture. The concentration of the porogen and, consequently, the amount of pores is different at the bottom and in the upper part of the mixture.

3.3. Crosslinking Polymerization in the Presence of Substances Releasing Gases which Remain in the Resulting Hydrogel

The basis of this technique is a crosslinking polymerization in the presence of components that release gases. Subsequently, the bubbles of gas evolve in the liquid polymerization mixture and are fixed in the polymerizing mass giving rise to the pores. Usually, it is necessary to add foam stabilizer into the polymerization mixture. Its function is to stabilize the foam so that the polymerization can be finished at the foam stage.

For example, superporous poly(*N*-isopropylacrylamide) or polyacrylamide hydrogels [10, 15] with the pore size 100 μm and larger was prepared by polymerization of the monomers, crosslinker (*N,N′*- methylenebisacrylamide), initiator (ammonium peroxosulfate), butane-1,4-diamine hydrochloride, stabilizing agent and sodium hydrogencarbonate. $NaHCO_3$ had two functions: First, it reacted with an acid to generate CO_2 bubbles, which were essential in making the foam. Second, it increased pH of the solution and accelerates the polymerization. A typical morphology of this porous hydrogel is shown in figure 6. [14].

In this manner it is possible to prepare porous hydrogels with communicating or non-communicating pores, whereas the ratio of disconnecting pores is similar or higher compared with the preparation techniques mentioned above. A disadvantage of this method is complicated procedure in preparation. On the other hand, using this technique porous hydrogels can be prepared from the substances whose monomers are not available, for example chitosan and glycol chitosan [16, 17]. The porous structure was obtained by the addition of glyoxal as a crosslinker and sodium hydrogencarbonate to a solution of chitosan in acetic acid. Carbon dioxide was evolved in the reaction of sodium hydrogencarbonate with acetic acid.

The block crosslinking polymerization in the presence of a small amount of low-temperature boiling solvent can also be considered as this technique of preparation. The gas bubbles are produced by the solvent, boiling due to the heat released at the gelation point [40]. However, this method is not commonly used because of a very low reproducibility of the final porous material and predominantly noncommunicating pores.

3.4. Freeze-Sublimation of the Hydrogel Swollen in Water (Lyophilization of Swollen Hydrogel)

The water-swollen hydrogel is frozen and water is sublimed under vacuum at low temperature. This technique affords only limited pore size regulation. Also because of the wide distribution of pore sizes [4,41], this technique is one of the least frequently used. Nevertheless, similarly like in the method 3.3., this technique enables the preparation of porous structures from the polymers whose monomers are not available. A typical structure of the material prepared in this manner is shown in figure 7. [4]. Apparently, the pore walls are very thin, even thinner than those of hydrogels prepared by precipitation polymerization (3.1.). A promising biodegradable porous hydrogel, poly(sodium alginate), was prepared by a three-step technique [4]: A dilute alginate solution (2%) was transformed to a gel in homogenizer to form a crosslinked hydrogel, then frozen and finally dried by lyophilization.

4. Characterization of Porous Hydrogels

4.1. Mercury Porosimetry

The method, invented in 1945 by H.L. Ritter and L.C. Drake [42], is based on the property of mercury that does not wet the surface of solid materials. During the measurement, the sample is placed in an evacuated cell, and mercury is then transferred into the sample cell under vacuum. Subsequently the pressure is applied to force mercury to penetrate the sample. During the measurement, the applied pressure and intruded volume of mercury are recorded. As a result, an intrusion-extrusion curve is obtained. The parameters describing the pore structure of the sample can be calculated from the data.

With the approximation that all pores are of cylindrical shape, the Washburn equation was deduced: [43]

$$p\,r = -2\,\gamma \cos \Theta$$

where p is the applied presure, r is pore radius, Θ is the contact angle of mercury on the surface of solid sample and γ is the surface tension of mercury.

This equation is based on a simplified principle, however, real measurements are considerably influenced by many other factors. The penetration of mercury into the pores of a sample is a process affected by the size, quantity and shape of the pores, but also by the size of the sample. During the measurement, the adjusted rate of pressure increase is important; the time of attaining equilibrium is observed. The first part of an intrusion-extrusion curve depicts the pressure increase. With increasing pressure mercury gradually penetrates into smaller pores of the sample. At the limiting pressure 1000 MPa, mercury penetrates into the 2 nm pores. The second part of the curve depicts extrusion of mercury from the sample. The amount of mercury that remains in the pores predicts the pore character.

According to the shape of the intrusion-extrusion curve several shapes can be distinquished:

a. Cylindrical pores are characteristic by a rapid penetration part of the curve. The extrusion part has a similar shape to the intrusion one. A negligible amount of intruded mercury remains in the pores.
b. The penetration curve obtained for conical pores shows a slow increase caused by gradually decreasing pore diameter. Again, the extrusion curve follows the intrusion one. Conical pores are the most common in materials of natural origin.
c. Slot pores having a shape of the space between two parallel plates are typical by a rapid increase of intrusion curve at higher pressures. A large amount of mercury remains in the pores after extrusion.
d. Ink-bottle pores have narrow entrance to the wider internal space. After the exclusion almost all mercury remains in the pores.

The sample porosity is calculated from the mass of mercury intruded to the pores under the highest pressure used. Mercury fills the whole void space except the pores smaller than 2 nm. Using the mathematic apparatus corresponding to the particular shape of the pores, the specific surface area can be estimated from the total pore volume and pore diameters. For example, for the pores of conical shape the surface is different than for the pores of the same size and volume being the same. From a comparison of the estimated surface area with the result of exact measurement by BET technique (see below), the predominant pore shape can be found. The most often presented result of mercury porosimetry is a pore size distribution curve depicting a semilogarithmic dependence of cumulative pore volume on the pore diameter interval. The results are also interpretable as the dependence of pore occurence per the interval of their diameter. The maximum of the curve depicting this dependence coresponds to the most frequent pore radius.

The main disadvantage of mercury porosimetry is the possibility of measuring only dry samples. Particularly for the hydrogels with high swelling degrees, the difference between pore volume and shape in the dry and swollen state is substantial. A high pressure is used in the measurements, which can also compress the sample.

4.2. BET Surfeace Area Measurements

The technique invented in 1938 by S. Brunauer, P.H. Emmett and E.J. Teller [44] is the most frequent method for determination of specific surface of porous materials. The measurement is based on physical adsorption of gas on the sample surface. The amount of gas can be defined by the Langmuir isotherm assuming a monolayer of gas molecules on the homogenous surface which are not mutually affected.

$$a = q\,K\,p/(1 + K\,p)$$

where a is the amount of the gas adsorbed in 1 g of sample (mol/g), q is the amount active centers in 1 g of sample (mol/g), p is partial pressure of the gas and K is equilibrium coefficient of adsorption.

For a sample covered by gas molecules in more than one layer, the BET isotherm is applicable:

$$\frac{p}{a\,(p^0 - p)} = \frac{1}{a_m C} + \frac{C - 1}{a_m C} \cdot \frac{p}{p^0}$$

where p^0 is tension of saturated vapour at particular temperature, a_m is the of gas adsorbed in 1 g of sample in a monolayer (mol/g) and C is a constant including adsorption and condensation heat.

The adsorption isotherm depicts the dependence of the amount of adsorbed gas on the pressure at constant temperature. From the measured isotherm the amount of gas adsorbed in monolayer is determined; the area of sample covered by one molecule of gas:

$$A_{SP} = N_A \,.\, a_m \,.\, \sigma$$

where A_{SP} is the specific surface area, N_A is the Avogadro constant and σ is the area of the sample covered by one molecule of the adsorbing gas. The measured values of specific surface area depend on the gas adsorbed – lower values are obtained with bigger molecules. Usually nitrogen or argon is used at low tempreatures.

Another method used for determination of specific surface area is adsorption of a chemical compound from an organic solvent (decrease is determinated calorimetrically) or adsorption of isotopically labelled compound. However, the molecules of these compounds are usually considerably larger, they do not get into the smallest pores and the measured values are lower compared with those obtained using the BET technique.

4.3. Scanning Electron Microscopy (SEM)

The SEM image is a result of the interaction of the sample with electron beam. Many factors such as electron energy, sample density, atomic number of elements and, clearly, topography of the sample surface affect this interaction. Elastic and non-elastic interactions of electrons with the atoms of sample generate secondary electrons, Auger and back-scattered electrons, continuum and characteristic X-rays and fluorescence. Usually the secondary electrons are used for the SEM purpose, other products can also bring important information about the sample and they are used in other spectroscopic techniques.

An advantage of SEM is high depth of sharpness providing information about structures at various distances from the level of scanning but, on the other hand, it precludes simple measurement of objects in the 2D depiction. For this purpose some special techniques must be used [45, 46].

To obtain information about the morphology of a porous hydrogel it is necessary to know the inner structure. The proper preparation technique based on fracture or cutting provides a new surface of the sample representing its morphology without any secondary changes. The incident electrons bring negative charge on the sample which has to be removed. To avoid a repulsive reaction of electron beam the sample surface is usually covered with a thin layer of gold. A main disadvantage of SEM is that the observation proceeds under vacuum. During drying, secondary changes of the sample structure can appear. This effect can bring limitations, especially for highly swollen porous hydrogels.

A modification of scanning electron microscopy - LVSEM (low-vacuum SEM) - works with a two-chamber system where the first high-vacuum detecting chamber is separated from the second low-vacuum chamber by the shade with a small hole. Although the magnification is lower compared with SEM the sample could not be covered with gold. [47, 48]. Environmental scanning electron microscopy (ESEM) proceeds under natural conditions. Wet samples without any pretreatment are observed in a small chamber with water vapour [47]. Another modification is AQUASEM making it possible to observe the samples under the wet conditions as well.

4.4. Confocal Microscopy

The source of light is a laser beam. Confocal microscope provides extremely sharp, contrast, highly informative images with a high resolution. The depth of sharpness is always very small and the sample structure above and underneath the focusation does not affect the quality of image.

During the preparation the hydrogel is stained with a fluorochrome and the laser beam is focused on a particular layer intersecting the sample as an imaginary cut. The sample material at this layer is illuminated with a laser beam and the fluorochrome emits a visible light. The obtained image is recorded with computer. By sequential focusing it is possible to achieve sharp images from various depths of the sample. This is a crucial difference from conventional fluorescence microscopy where the fluorochrome shines at all levels and the depth is not distinguishable.

Reaching the lower distance between the scanned layers than the depth of sharpness it is possible to make a very detailed 3D reconstruction of the sample structure. Together with the sophisticated computer processing, a three-dimensional depiction of the porous sample can be made at various depths or even with perspective.

To reach higher sharpness of the image deconvolution, i.e., mathematic correction of depiction can be used.

The characterisation of macroporous hydrogels by confocal microscope and by SEM was compared using the HEMA/MANa copolymers [5].

If we then intend to perform 3D reconstruction of an image, the distances between individual scan planes should be always smaller than the depth of sharpness of individual images. After taking images of more planes (commonly 30-60) the individual images are stacked along the z-axis (arranged above each other in the order of taking them). In this way, a specific value of brightness is obtained for every point of the observed object (for a given fluorochrome). The direction and angle of the view can be changed at random. The object can be imaged from any side (angle of view); the object can be even imaged from side or from underneath (all image data from all virtual cuts are saved in PC). All that is calculated again for a new chosen direction and angle of view (based on the image data from all imaged layers) and, subsequently, a view of the object from the chosen direction and angle is generated. At present software permits also 3D perspective imaging (the nearer parts of the object appear bigger than the more distant parts). Software simulation of partial transparency of the object is necessary so that it can be distinguished what is more up or down.

For attaining a higher sharpness of image, deconvolution can be also used. This is a mathematical method of object correction. If the degree of distortion is known (it can be described mathematically as convolution using functions of point coverage) then the image can be subjected to deconvolution. The image after deconvolution thus resembles more the real shape of the object than that obtained with microscope and thus it becomes sharper, with a lower noise and higher resolution. In practice the deconvolution is performed as follows. Based on the images of a small latex ball (its shape is known) taken with a specific microscope, the degree of distortion of the given optical system is determined. Subsequently, the obtained image is recalculated to image the ball as a ball. The correction is then used also for other objects. For example, confocal microscopy was used for characterization of macroporous HEMA/MANa copolymer hydrogels and in comparison with electron microscopy [3].

Both, electron and confocal microscopy are indirect methods of hydrogel morphology study and the micrographs obtained by them can be markedly different. The confocal micrograph of the same hydrogel as on figure 3 is shown on figure 8.

4.5. Diffusion Properties

Particularly for porous hydrogels employed as drug carriers the diffusion parameters play a key role. Several experimental techniques were used for assessment of these properties. For example Lévesque et al. [3] used the following methodology for determination of the diffusion parameter (D_{eff}): The hydrogel sample was immersed into a solution of the studied protein until equilibrium was achieved. Subsequently, the sample was immersed into water and the dependence of the released protein concentration on time was investigated. The diffusion parameter was calculated according to:

$$M_t/M = 4(D_{eff}t/\pi r^2)^{1/2},$$

where M_t is the amount of released protein at time t, M is the starting amount of protein in the sample, D_{eff} is the diffusion parameter and r is diameter of the cilindric sample.

Since D_0 is the diffusion paramater of the investigated substance in water then:

$$D_{eff} = D_0/\tau,$$

where τ is tortuosity.

Tortuosity determines the prolongation of the trajectory of molecule traveling in the hydrogel, i.e., trajectory of the diffusing molecule is not straight as in liquid but it is longer due to the hydrogel walls. As the value of the hydrogel tortuosity is higher, the molecule diffuses through it longer.

4.6. Mechanical Properties

The employment of hydrogels as implants in the human body requires similar mechanical properties as the surrounding tissue. For example, for application of porous hydrogel in the central nervous system, it should be as soft as possible. On the other hand, it must sustain necessary manipulation. In contrast, the hydrogels used for cultivation of bone cells should be solid and hard.

Woerly et al. [33] used a low-amplitude oscillatory shear measurement on a stress rheometer. A special cylindrical cell with a serrated bottom plate was designed to run measurements in a liquid environment. The sample of hydrogel (or neural tissue for comparison) was cut into disks of 15 mm in diameter and 1.3 mm in thickness and then introduced in the cylindrical cell and slightly compressed by upper serrated plate to create a fixed gap. A gap of 1.1 mm was chosen to compare the mechanical properties of the hydrogel and the neural tissue. In addition, successive oscillatory shear measurements were carried out on the gel subjected to comparison and decompression cycles. Starting at 1.1 mm, the gel was smoothly compressed to 1.05, 1.0, 0.95 and 0.9 mm and subjected to oscillatory shear for each gap. After each compression the sample was left for 5 – 15 min to ensure osmotic equilibrium between the solution and the gel prior to testing. The cell was filled with a saline solution prior to testing and all experiments were conducted at 37 °C. To avoid evaporation of the solution the cylinder was isolated with a teflon cover. The frequency (ω) was varied from 0.01 to 100 rad s^{-1}. Stress sweep measurements were first conducted to delimit the region of linear viscoelasticity, that is the stress, τ^*, is proportional to the strain, γ^*. Depending on the frequency range, the stress was adjusted so that the rheological behavior was linear viscoelastic. Care was taken to impose adequate stresses so that the resulting strains were small enough and did not break the structure of the gel. Oscillatory shear tests were carried out from higher frequency measurements. Under small harmonic strains, the ratio of the stress to the strain is independent of the imposed strain [33]:

$$G^*(\omega) = \tau^*(t)/\gamma^*(t) = G'(\omega) + \mathrm{i}G''(\omega),$$

which means that stress is a linear function of strain (domain of linear viscoelasticity). $G^*(\omega)$ is the complex shear modulus; $G'(\omega)$ is called the storage modulus and represents the elastic contribution and $G''(\omega)$ the loss modulus and represents the viscous contribution to the viscoelastic behavior of the material. The complex dynamic viscosity is obtained by the ratio:

$$\eta^*(\omega) = G^*(\omega)/\mathrm{i}\omega.$$

5. Modification of Porous Hydrogels

For some purposes the modification of porous hydrogels is advantageous. For example in some biomedical applications their modification with biologically active proteins is useful. Synthetic hydrogels without any charge can be modified with proteins by nonspecific interactions [37]; however, the amount of protein is usually small and interaction is not always strong enough. Another type of immobilization of protein by nonspecific interactions consists in the use of a protein-synthetic polymer composite [49]. A considerably stronger interaction is obtained using charged hydrogels. The charge choice depends on the protein to be immobilized. Generally, proteins with pI > 7 are easily bonded to hydrogel with the negative charge and protein with pI < 7 to those with positive charge at pH 7 [37]. For example, the amount of immobilized albumin (pI 4.9) on the EOEMA-MOETACl copolymer is 300 times higher than when nonspecific immobilization is used. On the other hand, the HEMA – MANa copolymer HEMA – MANa copolymer is suitable for avidin (p*I* 10.05) [37]. The amount of immobilized protein is affected also by pH of the medium and its ionic strength. pH determines the charge amount in protein; increasing of protein charge leads to an increased amount of immobilized protein and a structure similar to polyelectrolyte complexes arises. The effect of ionic strength is similar; more low-molecular-weight ionts in the medium leads to a smaller amount of charge in protein (and in hydrogel as well) and the bond between the protein and hydrogel is weaker.

For example, oligopeptides and aminosugars [50, 51] can be covalently bonded to the copolymers consisting of MA by the methods described in the review [52].

Deposition of proteins by matrix-assisted pulsed-laser evaporation [53] makes it possible to prepare protein films of controlled nanometer thickness on the surface of synthetic substrates.The method is of interest for various applications, such as preparation of active layers in biosensors or biocompatible coating of medical materials.

6. Author´s Experience with Porous Hydrogels Prepared in the Presence of Porogen Particles

6.1. Porous Hydrogels (According to 3.2.) for Tissue Engineering

Tissue engineering uses polymer scaffolds seeded with cells to repair damaged tissues and organs. The scaffolds should support cell adhesion and functions leading to the formation of new tissue. Synthetic hydrogels are suitable materials for many applications owing to their mechanical properties similar to extracellular matrix and soft tissues [54, 55]. The hydrogel structure containing a considerable amount of water allows easy transport of gases, ions, and low-molecular-weight nutrients and waste products. Hydrogels can be prepared in various shapes and porosities, with mechanical properties controlled by crosslinking or by copolymer composition. Their softness facilitates low-invasive implantation [56].

The research and development of hydrogels has in our Institute and especially in our Department a long tradition. We employ various copolymers of hydrophilic methacrylates first described by Wichterle and Lim in the early sixties [57]. In the area of porous hydrogels we started with poly(2-hydroxyethyl methacrylate) assuming its ability to control pore size and verify methods of characterization (6.2.) [5].

As the second step, we studied the influence of charged functional groups incorporated in polymer network on the growth of seeded cells. Comparing of positively or negatively charged porous hydrogels with neutral ones, we obtained the best results with the positively charged materials [36]. Further aim of our research was a model study of protein adsorption on the surface of porous hydrogels [37]. The reason was that the original surface structure or immobilized low-molecular-weihht substances in contact with physiological fluids or cell culture media can be easily overlayered by large proteins with low biological activity (such as albumin). Hence, the accessibility of above structures is considerably reduced.

The growth of bone marrow stromal cells was observed *in vitro* in macroporous hydrogels based on HEMA copolymers with different electric charges. The prepared macroporous hydrogels with communicating pores are materials formed by a three-dimensional polymer network with 10–100 μm pores. They have a large surface available for cellular ingrowth and are strongly hydrated in physiological solution. They are in many ways similar to the natural environment in growing nervous tissue and are capable of providing mechanical support to ingrowing cells and axons. Their chemical (composition, charge) and physical properties (porosity, pore size, surface and mechanical properties) can be tailored to a specific use [38].

For many purposes, it is appropriate to modify biomaterials so that they can degrade after some time in contact with living tissue to water-soluble substances which can then be eliminated from the organism. The preparation of degradable porous hydrogels was described in [1]. Macroporous hydrogels based on 2-hydroxyethyl methacrylate, 2-ethoxyethyl methacrylate and *N*-(2-hydroxypropyl)methacrylamide, methacrylic acid and [2-(methacryloyloxy)ethyl]trimethylammonium chloride crosslinked with *N,O*-dimethacryloylhydroxylamine were prepared. Hydrogels were degraded in a buffer at pH 7.4. Completely water-soluble polymers were obtained over time periods ranging from 2 to 40 days. The degradation process was followed gravimetrically and by optical and electron microscopy. *In vivo* biological tests with hydrogels based on copolymers of 2-ethoxyethyl methacrylate/*N*-(2-hydroxypropyl)methacrylamide were performed.

Several types of porous hydrogels based on HEMA were implanted inside a spinal cord hemisection cavity at the Th8 level in a rat. The spinal cords were processed 1 month after the implantation and histologically evaluated. All hydrogels adhered well to the surrounding spinal cord tissue. They formed a firm bridge across the hemisection cavity. It was shown that HEMA-based hydrogels could bridge a spinal cord injury. Further, positively charged functional groups promote connective tissue infiltration and both positive and negative charges induce axonal regeneration inside a hydrogel bridge [58].

6.2. Characterization of the Porous Hydrogels Prepared According to 3.2

The specific number, volume and surface of the pores (i.e. the number, volume and surface of the pores per 1 cm^3), pore diameter and single-pore volume can be easily determined without any instrumentation. The method is based on the known values of the mean diameter of sodium chloride particles used in the preparation of hydrogel, volume of water in the equilibrium-swollen hydrogel and diameters of the hydrogel sample [5]. The determination is based on the simplifying assumption of spherical shape of NaCl particles with the same volume as the total volume of real particles and Gauss distribution of particle sizes during fractionation.

The number of pores in the macroporous hydrogel in 1 cm^3 of sample, i.e. the number of NaCl particles (n) is:

$$n = m_{\mathrm{NaCl}}/[\rho_{\mathrm{NaCl}}.(4/3)\pi.(d/2)^3.V_{\mathrm{H}}],$$

where ρ_{NaCl} is the density of solid sodium chloride (2.16 g.cm^3), m_{NaCl} is total weight of NaCl used in sample preparation, V_{H} is volume of equilibrium swollen sample after NaCl removal by washing out and d is the average size of the fractionated sodium chloride particles. For example, NaCl particles fractionated between sieves with mesh 30, 50 and 90 μm were used for the preparation of HEMA-based macroporous hydrogels [5].

Total pore volume in 1 cm^3 of the hydrogel, i.e. the volume of water in all pores (V_V) is:

$$V_V = [V_H - m_H/(\rho_p.Z_V)]/V_H,$$

where ρ_p is the density of dry polymer (commonly 1.2 $g.cm^3$) and Z_V is the volume fraction of dry polymer in equilibrium-swollen hydrogel.

The average volume of one pore is:

$$V_1 = V_V/n.$$

Diameter of one pore in macroporous hydrogel is:

$$d_H = 2[3V_1/(4\pi)]^{1/3}$$

and total surface area of all pores in 1 cm^3 of macroporous hydrogel is:

$$S = 4\pi(d_H/2)^2.n\ .$$

This method of porous hydrogel characterization has important advantages. It is applicable to the sample in swollen state without any drying like in the other techniques. Also porous hydrogels with non-communicating pores can be characterized. The only limitation is the technique of preparation using soluble particles as porogen. The difference between the determined pore diameter and the size of porogen particles increases with increasing hydrophilicity of the porous hydrogel. An example of this difference for macroporous hydrogel HEMA/MANa is shown in figure 9.

6.3. Characterization of through-Flow Properties of the Hydrogels with Communicating Pores

For some applications of porous hydrogels it is necessary to characterize their through-flow properties. The hydrogel through-flow behavior also predicts interconnections of the pores. The flow rate can be easily determined using the apparatus shown in figure 10. [40]. The 8 mm wide and 12 mm long cilinder of equilibrium swollen hydrogel in saline was poured into a glass column with inner diameter 7.6 mm. The volume above the sample was filled with saline. The volume of solution that flows through the sample (expressed in drops of average volume 24.5 mm^3) was measured as a function of time and hydrostatic pressure. The specific flow-rate, i.e. the flow-rate through the column of unit length (1 mm) and unit bottom area (1 mm^2) , was determinedfrom the equation:

$$v_{sp} = [(n\ .\ 24.5\ .\ l)/(t\ .\ \mathrm{S})]\ .\ (S_0/l_0) \qquad (\mathrm{mm^3 s^{-1}}),$$

where v_{sp} (mm^3s^{-1}) is flow rate, n is the number of drops (average drop volume 24,5 mm^3), l is the length of the sample (mm), t is the time (s), S is the bottom area of the sample column, S_0 is the unit area (1 mm^2) and l_0 is the unit length (1 mm).

7. Perspective

All the teams of researchers are mainly focused on the one method of macroporous (superporous) hydrogel preparation, but a comparison of hydrogels of the same chemical structure prepared by various methods has not yet been published. This article could help to choose the proper method of preparation of a porous hydrogel for a given purpose. Also the number of papers concerning the hydrogel surface modification is very low, whereas these modifications can significantly extend the field of the hydrogel usage.

One of the topics of future research could be the application of new technique of covering just of the pore walls in porous materials with a monolayer of selected proteins (collagen and subsequently fibronectin or laminin). The procedure is described in literature [56, 59]. Athin layer of suitable proteins on the whole surface of porous scaffold, including pore walls, will improve its attraction to the seeding cells. The method opens the wide spread area of employing various bioactive motifs for selected cell types.

Another way of development of porous hydrogels (including the degradable ones) is the preparation of "smart" materials which are degradable enzymatically or hydrolytically, with induction period.

Porous hydrogels seem to be interesting materials not only as supporting scaffolds for tissue engineering but their similarity to living tissue and controllable properties give them potential to play a more active role in various medical applications such as artificial substitution of damaged tissue.

As it was mentioned in paragraph 2, bunchs of nanofibers are also one of type of porous hydrogel. From the tissue engineering point of view, such materials seem to be even better than other porous hydrogels, cells proliferation is very good as it is shown in figure 11.

Acknowledgments

The study was supported by grants from the Academy of Sciences of the Czech Republic, Projects No. 1QS400500558 and MSMT 1M0538.

8. References

[1] Pradny, M; Michalek, J; Lesny, P; Hejcl, A; Vacik, J; Slouf, M; Sykova, E. E. Macroporous hydrogels based on 2-hydroxyethyl methacrylate. Part 4. Hydrolitically degradable materials. *J. Mater. Sci. Mater. Med.* 2006, 17, 1357 - 1366.

[2] Mikhalovska, LI; Gun'ko, VM; Turov, VV; Zarko, VI; James, SL; Vadgama, P; Tomlins, PE; Mighalovsky, SV. Characterization of the nanoporous structure of collagen-glycosaminoglycan hydrogels by freezing-out of bulk and bound water. *Biomater.* 2006, 27, 3599 - 3607.

[3] Lévesque, SG; Lim, RM; Shoichet, MS. Macroporous interconnected dextran scaffolds of controled porosity for tisue-engineering applications. *Biomaterials.* 2005, 26, 7436 - 7446

[4] Shapiro, L; Cohen, S. Novel alginate sponges for cell culture and transplantation. *Biomater.* 1997, 18, 583 - 590

[5] Pradny, M; Lesny, P; Fiala, J; Vacik, J.; Slouf, M; Michalek, J; Sykova, E. Macro porous hydrogels based on 2-hydroxyethyl methacrylate. Part 1. Copolymers of 2-hydroxyethyl methacrylate with methacrylic acid. Collect. *Czech. Chem. Commun.* 2003, 68, 812 - 822.

[6] Liu, Q; Hedberg, EL; Bahulekar, R; Meszlenyi, RK; Mikos, AG. Preparation of macroporous poly(2-hydroxyethyl methacrylate) hydrogels by enhanced phase separation. *Biomater.* 2000, 21, 2163 - 2169.

[7] Chen, F-M; Zhao, Y-M; Sun, H-H; Jin, T; Wang, Q-T; Zhou, W; Wu, Z-F; Jin, Y. Novel glycidyl methacrylated dextran (Dex-GMA)/gelatin hydrogel scaffolds containing microspheres loaded with bone morphogenetic proteins: Formulation and characteristics. *J. Controlled Release.* 2007, 118, 65 - 77.

[8] Courtney, JM; Murphy, SM; Robertson, LM; Ryan, CJ; Tighe, BJ. Examination of particulate macroporous hydrogels in an extracorporeal rat hemoperfusion model. *J. Biomater. Sci., Polym. Ed.,* 1999, 10, 1063-1077.

[9] Chen, J; Park, H; Park, K. Superporous hydrogels: fast responsive hydrogel systems. *Polym. Mater. Sci. Eng.,* 1998, 79, 236 - 237.

[10] Chen, J; Park, K. Superporous hydrogels: Fast responsive hydrogel systems. *J. Macromol. Sci. Pure Appl. Chem.,* 1999, 36, 917 - 930.

[11] Chen, J; Park, H; Park, K. Synthesis of superporous hydrogels: Hydrogels with fast swelling and superabsorbent properties. *J. Biomed. Mater. Res.,* 1999, 44, 53 - 62.

[12] Kim, D; Park, K. Swelling and mechanical properties of superporous hydrogels of poly(acrylamide-co-acrylic acid)/polyethylenimine interpenetrating polymer networks. *Polymer,* 2004, 45, 189 - 196.

[13] Diez-Pena, E; Quijada-Garrido, I; Barrales-Rienda, JM. Acrylic hydrogels with special properties. *Rev. Plast. Mod.* 2003, 85, 466 - 476.

[14] Tang, C; Yin, C; Pei, Y; Zhang, M; Wu, L. New superporous hydrogels composites based on aqueous Carbopol solution (SPHCcs): synthesis, characterization and in vitro bioadhesive force studies. *Eur. Polym. J.,* 2005, 41, 557 - 562.

[15] Dorkoosh, FA; Stokkel, MPM; Blok, D; Borchard, G; Rafiee-Tehrani, M; Verhoef, JC; Junginger, HE. Feasibility study on the retention of superporous hydrogel composite polymer in the intestinal tract of man using scintigraphy. *J. Controlled Release,* 2004, 99, 199 - 206.

[16] Park, H; Park, K; Kim, D. Preparation and swelling behavior of chitosan-based superporous hydrogels for gastric retention application. *J. Biomed. Mater. Res.* 2006, 76A, 144 - 150.

[17] Park, H; Kim, D. Swelling and mechanical properties of glycol chitosan/poly(vinyl alcohol) IPN-type superporous hydrogels. *J. Biomed. Mater. Res.* 2006, 78A, 662 - 667.

[18] Bajpai, SK; Bajpai, M; Sharma, L. Investigation of water uptake behavior and mechanical properties of superporous hydrogels. *J. Macromol. Sci. Pure Appl. Chem.,* 2006, 43, 507 - 524.

[19] Shi, QH; Zhou, X; Sun, Y. A novel superporous agarose medium for high-speed protein chromatography. *Biotechnol. Bioeng.,* 2005, 92, 643 - 651.

[20] Savina, IN.; Mattiasson, B; Galaev, IY. Graft polymerization of acrylic acid onto macroporous polyacrylamide gel (cryogel) initiated by potassium diperiodatocuprate. *Polymer,* 2005, 46, 9596 - 9603.

[21] Kroupova, J; Horak, D; Pachernik, J; Dvorak, P; Slouf, M. Functional polymer hydrogels for embryonic stem cells support. *J. Biomed. Mater. Res. Appl. Biomater.,* 2006, 76B, 315 - 325.

[22] Sakamoto, M; Nakasu, M; Matsumoto, T; Okihana, H. Development of superporous hydroxyapatites and their examination with a culture of primary rat osteoblasts. *J. Biomed. Mater. Res.* A, 2007, 82, 238 - 242.

[23] Omidian, H; Rocca, JG; Park, K. Elastic, superporous hydrogel hybrids of polyacrylamide and sodium alginate. *Macromol. Biosci.,* 2006, 6, 703 - 710.

[24] Tang, C; Yin, L; Yu, J; Yin, C; Pei, Y. Swelling behavior and biocompatibility of carbopol-containing superporous hydrogel composites. *J. Appl. Polym. Sci.,* 2007, 104, 2785 - 2791.

[25] Yin, L; Fei, L; Cui, F; Tang, C; Yin, C. Superporous hydrogels containing poly(acrylic acid-co-acrylamide)/O-carboxymethyl chitosan interpenetrating polymer networks. *Biomater.,* 2007, 28, 1258 - 1266.

[26] Yin, L; Fei, L; Tang, C; Yin, C. Synthesis, characterization, mechanical properties and biocompatibility of interpenetrating polymer network-superporous hydrogel containing sodium alginate. *Polym. Int.,* 2007, 56, 1563 - 1571.

[27] Kim, D; Seo, K; Park, K. Polymer composition and acidification effects on the swelling and mechanical properties of poly(acrylamide-co-acrylic acid) superporous hydrogels. *J. Biomater. Sci., Polym. Ed.,* 2004, 15, 189 - 199.

[28] Gu, T; Zhou, W; Ma, G; Su, Z. Rigid gigaporous chromatographic media and their potential impact on downstream processing. *China Particol.,* 2005, 3, 349 - 353.

[29] Pradny, M; Martinova, L; Michalek, J; Fenclova, T; Krumbholcova, E. Electrospinning of the poly(2-hydroxyethyl methacrylate) and its copolymers with 2-ethoxyethyl methacrylate. *Centr. Eur. J. Chem.* 2007, 5, 779 - 792.

[30] Fakirov, S; Lin, RJT; Bhattacharyya, D. Nano- and microfibrillar composites from polymer blends: manufacturing, properties and applications. *Polym. Prepr. Am Chem. Soc.Div. Polym. Chem.,* 2007, 48, 905 - 906

[31] Rehakova, M; Martinova, L. unpublished data

[32] Han, SO; Son, WK; Youk, JH; Lee, TS; Park, WH. Ultrafine porous fibers electrospun from cellulose triacetate. *Mater. Lett.* 2005, 59, 2998 - 3001

[33] Woerly, S; Pinet, E; De Robertis, L; Bousmina, M; Laroche, G; Roitback, T; Vargova, L; Sykova, E. Heterogenous HPMA hydrogels for tisue repair and axonal regeneration in the injured spinal cord. *J. Biomater. Sci., Polym. Ed.* 1998, 9, 681 - 711.

[34] Sirc, J; Bosakova, Z; Coufal, P; Michalek, J; Pradny, M; Hobzova, R; Hradil J. Morphological and chromatographic characterization of molecularly imprinted monolithic columns. *e-Polym.* 2007, 117.

[35] Omidian, H; Rocca, JG; Park, K. Advances in superporous hydrogels. *J. Controlled release*. 2005, 102, 3 - 12.

[36] Pradny, M; Lesny, P; Smetana Jr., K; Vacik, J; Slouf M; Michalek, J; Sykova, E. Macroporous hydrogels based on 2-hydroxyethyl methacrylate. Part 2. Copolymers with positive and negative charges, polyelectrolyte complexes. *J. Mater. Sci. Mater. Med.* 2005, 16, 767 - 773.

[37] Michalek, J; Pradny, M; Artyukhov, A; Slouf, M; Vacik, J; Smetana Jr., K. Macroporous hydrogels based on 2-hydroxyethyl methacrylate. Part 3. Hydrogels as carriers for immobilization of proteins. *J. Mater. Sci. Mater. Med.* 2005, 16, 783 - 786.

[38] Lesny, P; Pradny, M; Jendelova, P; Michalek, J. Vacik, J; Sykova, E. Macroporous hydrogels based on 2-hydroxyethyl methacrylate. Part 4. Growth of rat bone marrow stromal cells in three dimensional hydrogels with positive and negative surface charges and in polyelectrolyte complexes. *J. Mater. Sci. Mater. Med.* 2006, 17, 829 - 833.

[39] Pradny, M. Unpublished data.

[40] Pradny, M; Michalek, J. Unpublished data.

[41] Kang, HW; Tabata, Y; Ikada, Y. Fabrication of porous gelatin scaffolds for tissue engineering. *Biomater.* 1999, 20, 1339 - 1344.

[42] Ritter, HL; Drake, LC. 1945. Pore-size distribution in porous materials. Pressure porosimeter and determination of complete macropore size distributions. *Ind. Eng. Chem. Anal.* 1945, 17, 782 - 786.

[43] E. W. Washburn. *Phys. Rev.* 1921, **17,** 273 - 281

[44] Brunauer, S; Emmett, PH; Teller, EJ. Adsorption of gases in multimolecular layers. *J. Am. Chem. Soc.* 1938, 60, 309 - 319.

[45] Minakuchi, H; Nakanishi, K; Soga, N; Ishizuka, N; Tanaka, N. Octadecylsilylated porous silica rods as separation media for reversed-phase liquid chromatography. *Anal. Chem.* 1996, 68, 3489 – 3501.

[46] Minakuchi, H; Ishizuka, N; Nakanishi, K; Soga, N; Tanaka, N. Performance of an octadecylsilylated continuous porous silica column in polypeptide separations. *J. Chromatogr.* 1998, 828, 83 - 90

[47] Danilatos, GD. Review and outline of environmental SEM at present. *J. Microsc.* 1991, 162, 391 - 402.

[48] Shah, J; Durkin, R. Experimental measurements of amplification in high pressure SEM. *Electron Microsc.,* 1992, 1, 109 - 110.

[49] Woerly, S; Maghami, G; Duncan, R; Subr, V; Ulbrich K. Synthetic polymer derivatives as substrata for neuronal adhesion and growth. *Brain Res. Bull.,* 1993, 30, 423 - 432.

[50] Plant, GW; Woerly, S; Harvey, AR. Hydrogels containing peptide or aminosugar sequences implanted into rat brain: Influence on cellular migration and axonal growth. *Exp. Neurol.,* 1997, 143, 287 - 299.

[51] Woerly, S; Laroche, G; Marchand, R; Pato, J; Subr, V; Ulbrich, K. Intracelebral implantation of hydrogen – coupled adhesion peptides: Tissue reaction. *J. Neural Transplant. Plast.* 1995, 3, 245 - 255.

[52] Hersel, U; Dahmen, C; Kessler, H. RGD modified polymers: biomaterials for stimulated cell adhesion and beyond. *Biomater.,* 2003, 24, 4385 - 4415.

[53] Jelinek, M; Remsa, J; Brynda, E; Houska, M; Kocourek, T. Thin layers of bovine serum albumin by matrix assisted pulsed laser evaporation. *Appl. Surf. Sci.* 2007, 254, 1240 - 1243.

[54] Lee, KY; Mooney, DJ. *Chem. Rev. Hydrogels for tissue engineering.* 2001, 101, 1869 - 1879.

[55] Drury, JL; Mooney, DJ. Hydrogels for tissue engineering: scaffold design variables and applications. *Biomater.* 2003, 24, 4337 - 4351.

[56] Brynda, E; Kysilka, J; Pradny, M; Lesny, P; Jendelova, P; Houska, M; Michalek, J; Sykova, E. Surface modification of hydrogels based on poly(2-hydroxyethyl methacrylate) with extracellular matrix proteins. *J. Mater. Sci., Mater. Med.,* submitted.

[57] Wichterle, O; Lim, D. Hydrophilic gels for biological use. *Nature.* 1960, 185, 117 - 118.

[58] Hejcl, A; Lesny, P; Sedy, J; Pradny, M; Michalek, J; Jendelova, P; Zamecnik J; Sykova, unpublished data

[59] Brynda, E; Riedel, T; Dyr, J; Houska, M; Bacakova, L; Filova, E; Chlupac, J; Lesny, P; Jendelova, P; Sykova, E. *Czech Patent Appl.* PV 2006-821 Method of preparation of controlled fibrin layers on solid surfaces.

In: Porous Media: Heat and Mass Transfer...
Editors: J. L. Acosta and A.F. Camacho
ISBN: 978-1-60692-437-2

Chapter 4

MONTE CARLO SIMULATIONS FOR THE STUDY OF DIFFUSION-LIMITED DRUG RELEASE FROM POROUS MATRICES

Rafael Villalobos[a] ***and Armando Domínguez***[b]

[a] División de Estudios de Posgrado (Tecnología Farmacéutica),
Facultad de Estudios Superiores Cuautitlán/UNAM,
Av. Primero de Mayo S/N, Cuautitlán Izcalli 54740,
Estado de México, México

[b] UAM-Iztapalapa, Depto. de Química,
Av. San Rafael Atlixco 186, Col.
Vicentina, 09340 Mexico City, Mexico

ABSTRACT

How fast can drug molecules escape from a controlled matrix-type release system? This important question is of both, scientific and practical importance, as increasing emphasis is placed on design considerations that can be addressed only if the physical chemistry of drug release is better understood. In this work, the labor performed by our group on the modeling of diffusion-controlled release systems is reviewed and updated with the inclusion of new results.

INTRODUCTION

This chapter is related to the problem of drug delivery from inert porous matrices. This kind of delivery plays an important role in pharmaceutics when a drug-controlled release is required [1]. It is commonly accepted that the structure of the dosage form influences the drug release kinetics and the final amount of released drug [2], but the question is how? This work focuses on the case of drug release from inert matrices, i.e. water-soluble drug is embedded in a finely dispersed state in an insoluble carrier material (excipient) and released by diffusion. The excipient is considered as an inert material, so the carcass does not change as a function of time, i.e. carcass degradation, dissolution, swelling, etc. is excluded. The model could also correlate when the drug is dispersed molecularly. The amounts of both constituents are crucial for the suitable design of pharmaceutical dosage forms since drug release from these devices is a percolating process in itself [3, 4]. If an initial drug concentration, C_0, is below the drug percolation threshold (e.g. for a cubic network this is equal to 0.3116 [5]) then a significant amount of drug can be trapped inside the excipient matrix and never be released. On the contrary, if the excipient concentration is beyond the excipient percolation threshold, then the tablet can break up, then leading to uncontrolled dissolution. Following [3], a drug delivery device can be depicted as a network of drug particles spanned through an excipient matrix. This network structure is what makes difficult to comprehend the delivery dynamics and mechanism since the drug diffusion process is occurring inside a low-dimensional space (LDS) rather than in a Euclidean one. In order to quantify the influence that the device structure properties notably those that can be controlled| by means of the initial drug spatial distribution, and the C_0 value, have on the drug release profile and on the final quantity of released drug, some studies have been performed by our group [6 - 8]. This problem was there studied via Monte Carlo computer simulations, where drug release was simulated as a diffusion-controlled process in low-dimensional media as cubic networks and Menger sponges. Here we revise the results obtained in these works.

The physical idea concerning the design of the drug-releasing device is that of a three-dimensional space in which drug and excipient grains of diverse sizes are randomly distributed throughout the available volume. The main aspects of this study deal with the following issues: (i) Menger sponges are used as a model of matrix dosing systems, (ii) to assess the influence of the structural properties of the porous medium on the distribution and extend of drug grains and on the transport (diffusion) properties of the delivery device, (iii) to consider the porosity of the drug delivery form as dynamic property that can be changing during the drug dissolution process; and (iv) to have the idea of employing a lattice model to represent a correlated porous substrate by means of the Dual Site Bond Model (DSBM).

SOME DRUG RELEASE KINETIC EQUATIONS

In spite of the complexity of the phenomena involved in drug release mechanisms, a number of equations have been proposed and are currently employed for describing experimental drug dissolution profiles, i.e. the amount of drug released as a function of time. The problem of the release rate from LDS (e.g. fractal clusters and porous networks) was first studied by Bunde et al. [9]. They specifically reported that the release rate follows a power-law. This has the form:

$$\frac{M_t}{M_\infty} = kt^n \tag{1}$$

where M_j is the amount of drug released at time j; k is an empirical determined parameter, and n depends on the system characteristics (geometry, topology, etc.) and drug release mechanism, that is why equation (1) has been widely used [10]. On the other hand, Kosmidis et al. [11-13] have observed that the Weibull function is the most appropriate equation to describe the entire duration of the drug release process, and the power-law can be considered as an approximation for short times. Furthermore, the Weibull function is consistent with the theoretical predictions that have been made under the frame of fractal kinetics [12]. The Weibull equation has the form:

$$\frac{M_t}{M_\infty} = 1 - \exp\left(-at^b\right) \tag{2}$$

where a and b are real constants. This model describes experimental dissolution data [13] quite well and a physical meaning of these constants is given in [12]. a is related to the specific surface of the dosage matrix form, and the exponent b is mainly related to mass transport characteristics of the device. It originates from the fact that a depletion zone is created gradually near the boundaries of the release device, and thus, the drug concentration in the device is not uniform. The present effort aims to quantify these quantities for the case of low-dimensional media by means of simulation methods, which are able to reproduce drug-release experiments and allow the virtual study of scenarios which are difficult, time consuming, or expensive to study by means of direct experimental work. This is an study of the drug release kinetics from matrix platforms by means of Monte Carlo simulations, which are based on the random walk model of Fickian diffusion with excluded volume interaction, i.e. it is assumed that moving particles act as hard spheres that collide with each other and with no possibility for a sphere to penetrate into another [14]. This seeks to clarify the influence that the device structural properties have on the drug release profile, and on the final amount of drug to be released. The main contributions are: (i) in previous works [9, 11] the porosity was kept at a constant value, whereas here this porosity follows a dynamical behavior, then allowing the study of the initial release step, which has been overlooked in cited works; and (ii) we consider the initial drug concentration and the drug spatial distribution as variables, both of them which have not been analyzed in earlier works.

The programs conceived to construct our solid porous structures, to characterize their structure and transport properties, and to simulate the drug release as a diffusion-controlled process are described next.

METHODS

Our main goal is to describe the escaping of particles from an inert release device (e.g. a tablet). For simplicity, the D-particles (i.e. drug particles, primary clusters of water-soluble drugs), and the E-particles (i.e. excipient particles or primary clusters of an inert material) are assumed to form binary powder mixtures, which can be compressed to form cubic tablets of volume L^3 with null porosity. During the compression process, different types of bonds (D-D, E-E, and D-E) between the particles are formed. Thus, the tablet can be imagined as a L^3 volume spanned in the form of a cubic lattice, where all lattice sites are occupied, thus forming clusters of substances D or E [2]. It is evident that if C_0 is varied; percolation phenomena can take place [3]. During the drug-release process, medicine should be first dissolved and then transported through the liquid phase. It is assumed that the dissolution step is instantaneous, thus the release should be controlled by the specific surface area of the dosage body and by the mass diffusion occurring in a LDS. In fact, when the release device is immersed in water, drug particles located at the device borders are quickly transported to the liquid phase and released to the surroundings; however, this material is quickly exhausted and more soluble matter has to be transported from far distances inside the releasing device. Evidently, this transport process becomes slow and inefficient at an early stage due to intricate diffusion mechanism within the matrix porous space, as this involves drug permeable (void volume) and drug impermeable (volume occupied by substance D or E) zones. Consequently, drug release kinetics has gotten anomalous characteristics [15].

To develop our study, two types of porous media models were used: (i) a fractal structure know as Menger sponge, and (ii) cubic lattices.

Menger Sponge

Frequently, real delivery systems are matrix platforms with fractal geometry, as reveled by techniques such as mercury porosimetry, where experimental data are interpreted to determine the fractal dimension of a porous body [16]. A tablet consisting of both a soluble and brittle drug (caffeine) and a non-swelling water insoluble polymer (ethyl cellulose), becomes porous during drug release. A leached and subsequently dried tablet represents, in a narrow range of resolution, a sponge-like structure. The distribution of pores corresponding to the original sites occupied by caffeine particles has been related to the fractal dimension, d_f, of monolithic devices, i.e. M (the mass) will grow as $M \propto r^{d_f}$, where r is it characteristic length scale and d_f is lower than the Euclidean dimension, $d = 3$ in this case. Furthermore, it has been found that the fractal dimension of the porous tablet resulting from a fixed mixing ratio depends on the particle size of the soluble substance and, in all cases $d_f \in (2.733, 2.838)$. Therefore, fractal structures are helpful models of drug delivery devices. The Menger sponge is characterized by $d_f = 2.727$, which is very close to the experimentally observed d_f values. This fractal structure has been used as a model of matrix dosing system [2]. Here, drug release is simulated by a diffusion-controlled process (random walk). Six different types of Menger sponges, i.e. six fractal porous structures having the same fractal dimension, but different values of the random walk dimension, $d_w \in [2.028, 2.998]$, were used as models of porous solid dosing systems, with the aim of elucidating the effect of the structural properties of the porous matrix on drug release behavior. Figure 1 shows the six used sponges, and figure 2 illustrates their generation procedure. It begins with a solid cube and engages an iterative process involving the removal of different parts of the initial cube. Let us label CMS_0 our initial cube. Then, CMS_1 will be generated by partitioning CMS_0 into 27 identical cubes and taking out the central one as well as the 6 ones that are located at the middle of each CMS_0 face. Next, CMS_2 is constructed by repeating this process for each one forming CMS_1. In this way, a nested sequence of configurations, CMS_i, can be produced. Each one of the Menger sponges built in this work are generated after performing three of these iterations (CMS_3). To build the diverse sponges, the seven sub-cubes removed in each iteration are (the position vectors (x, y, z) correspond to the axes shown in figure 2: $(1.3.2)$, $(1,3,3)$, $(2,1,3)$, $(2,2,1)$, $(2,2,2)$, $(2,3,1)$, $(3,2,2)$ for substrate AMS, $(1,1,3)$, $(1,2,2)$, $(1,3,1)$, $(2,3,1)$, $(3,2,3)$ for substrate BMS, $(1,2,2)$, $(2,1,2)$, $(2,2,1)$, $(2,2,2)$, $(2,2,3)$, $(2,3,2)$, $(3,2,2)$ for substrate CMS, $(1,2,2)$, $(1,3,1)$, $(1,3,2)$, $(2,2,3)$, $(3,1,2)$, $(3,1,3)$, $(3,3,2)$ for substrate DMS, $(1,1,1)$, $(1,2,2)$, $(2,1,2)$, $(2,2,1)$, $(2,2,3)$, $(2,3,2)$, $(3,2,2)$ for substrate EMS, and $(1,2,2)$, $(2,1,1)$, $(2,1,2)$, $(2,2,1)$, $(2,2,3)$, $(2,3,2)$, $(3,2,2)$ for substrate FMS.

In this case, the drug dissolution algorithm can be summarized as follows: (i) the sponge is generated from a lattice consisting of $27 \times 27 \times 27$ sites, and after the third iteration, each site is saturated with drug (if this site is forming part of the porous volume), or excipient (if the site forms part of the solid body). If drug is held in a site, then this site is called a drug particle, (ii) the drug particles are allowed to move throughout porous space according to the random walk model (i.e. according to a "blind ant" routine [17,18], and (iii) excluded volume interactions between the drug particles are assumed. Our particular routine considers a drug particle undergoing nearest-neighbor random site displacements in each one of the previously described sponges. The porous space of the Menger sponges is considered empty (there are no drug particles) and the sponge structure is repeated in all directions to render a homogeneous structure of a sufficiently large length. Then, a walking drug element is released at a randomly chosen site and the distance traveled is measured at time intervals corresponding to $t = 10k$, $k \in \{1,2,3,...\}$. The mean square distance $\langle r^2 \rangle$ traveled by the walking drug particle is average over 1000 different walks. In this case, the time unit corresponds to one Monte Carlo step (i.e. a number of walking attempts equal to total number of sites). In fractal bodies, diffusion becomes anomalous (i.e non-Fickian) and is consistent with the law $\langle r^2 \rangle = t^{2/d_W}$, where $d_W \in (2,3)$, thus implying a sub-diffusive behavior. Initially, the walking drug element "sees" a fractal structure as long as it is confined within a distance close to the size of the sponge (27 in this case); therefore diffusion is anomalous since $d_W \neq 3$. For longer times, however, the walking drug element traverses the entire sponge and diffusion becomes normal. The d_W value is determined in this work from the linear fit of $\log\langle r^2 \rangle$ against $\log t$ for $\langle r^2 \rangle \in (0.00, 182.23)$.

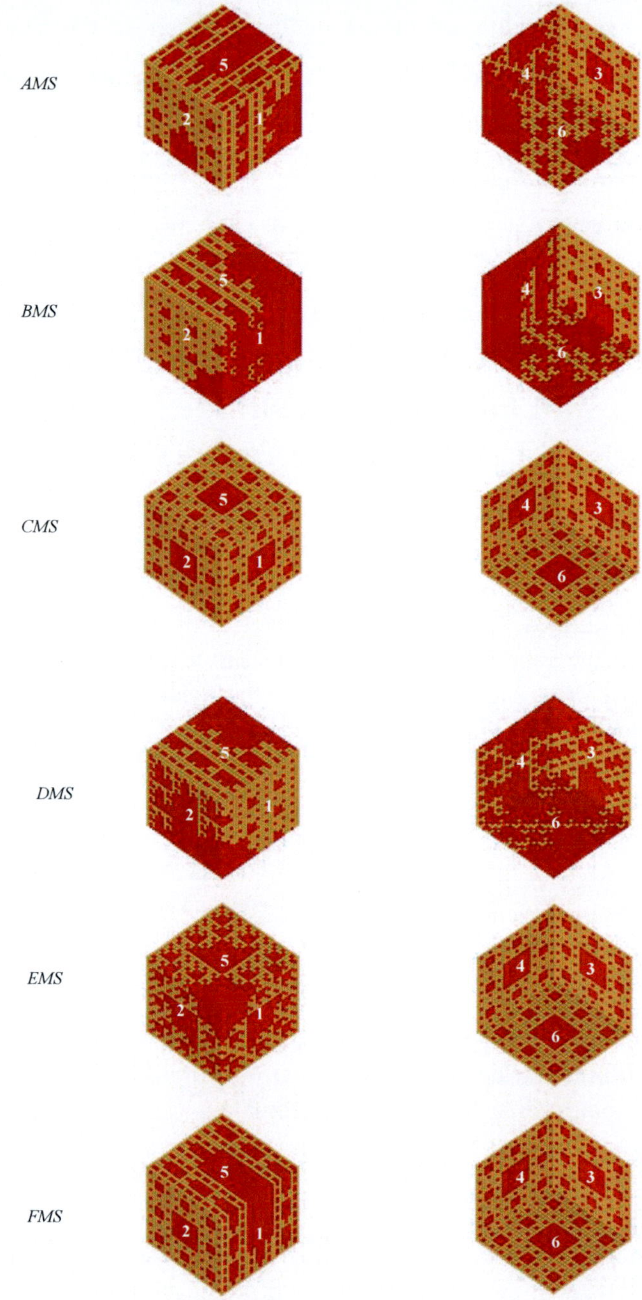

Figure 1. The Menger sponges used in this study. Darker sites correspond to drug particles, while clearer ones represent excipient particles. Numeric labels designate the different faces of each matrix platform.

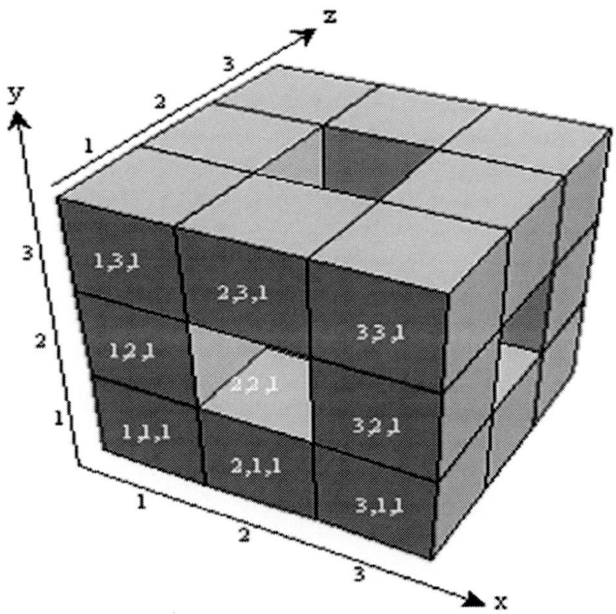

Figure 2. Illustration of the classical Menger sponge, and the axes of the vectors position.

Cubic Networks

Cubic lattices of sizes, $L \in \{10, 15, 20, 27, 35, 81\}$, and with various topologies are used as suitable models of drug release porous devices. In these models, the sites are occupied by one D-particle or one E-particle, according to a given C_0 value. C_0 ranges between the drug and excipient percolation thresholds. In all the simulations the initial porosity is taken as zero and the possibility of double occupancy is excluded. In this way, we can handle three-dimensional structures resembling real and plausible drug dosage forms. The drug release process is described as a diffusive escape of D-particles, equivalent to the ant in the labyrinth problem [2, 18]. The release process attempts to mimic the leaking of drug particles from a tablet, when this substrate (which is constitute by both, drug and excipient particles) is placed in contact with a solvent phase. The diffusion of drug particles from the substrate is modeled by selecting a drug particle at random and attempting to move it randomly to any one of its adjacent sites. If the chosen site is already occupied (by either another drug or excipient particle) the movement is rejected, but if the neighboring site is empty the movement is accepted. The drug particles can finally exit the tablet when they reach a site located at the sponge border. At this point, we assumed sink conditions, i.e. the drug is considered freely soluble in aqueous phase, so the medicine release is not limited by drug solubility. After each move (either successful or not) the time is incremented by a value equal to $1/N$, where N is the number of total drug particles remaining into the matrix [2]; this is a standard method to consider time in a Monte Carlo process [9,11]. The number of particles remaining in the matrix is monitored as a function of time, until only 10% of the drug particles remain inside the porous matrix. The release rate, dQ/dt, is monitored by counting the number of particles that diffuse into the leaking surface area (i.e. from the six faces of the sponge) in the interval between t and $t+1$. When a total of N_t D-particles have been chosen, this corresponds to a time step (Monte Carlo step, MCS). We average our results over 1000 realizations with different initial random configurations.

Cubic lattices were built by means of the Dual Site-Bond Model, DSBM. This is a Monte Carlo method already published and which so far has been applied to generate porous substrates and heterogeneous surfaces, at this respect please c.f. [19, 20] and references therein. This method is intended to mimic heterogeneous porous structures, i.e. pores of different sizes interconnected in the way of a simple cubic network, c.f. Figure 3, and can be summarized as follows. First, two real number distributions with a shared overlapping area, $\Omega \in [0,1]$, are set; one of them, F_b, comprises smaller numbers than the other, F_s. Figure 4 shows the uniform number distributions used in this work. The numbers are assigned to the sites (nodes) of the network by picking up randomly a real number, R, from the F_s distribution. Likewise, numbers are assigned to the bonds of the network by selecting a number at random from the F_b distribution function. Thus, at the end of the assignation process every site and bond in the network has an associated number identifying it. Afterwards, a spatial correlation among the sites of the network is disseminated by imposing the following condition: the associated number of every site should be greater or at least equal to anyone of its delimiting bonds. If $\Omega > 0$, consequently some sites could not satisfy the preceding condition; therefore, the spatial distribution of numbers over the elements of the network needs to be changed. To do this, the numbers of either two sites or two bonds selected at random are exchanged; the movement is accepted if after it the two selected elements fulfill the imposed correlation condition, if not the move is rejected. The movements continue until all the network sites satisfy the previous requirement. Thus the value of Ω determines the degree of correlation among the sites of the network. It has already been established [21] an empirical equation relating the value of the correlation length, l_0 (the average extent of sites patches having similar numbers), among the sites of a two-dimensional network to the overlapping Ω:

$$l_0 = \frac{2\Omega^2}{(1-\Omega)^2} \tag{3}$$

So, the parameter Ω is the predominant one that determines the topology of the porous space, i.e., the precise sequence, statistically expressed, in which the numerical labels of the void-elements distribute throughout the porous network. To clarify ideas, let us analyze the two limiting situations: (i) if $\Omega = 0$, then $l_0 = 0$, sites and bonds are not correlated at all and their associated numbers can be distributed completely at random, and (ii) if $\Omega = 1$ then $l_0 = \infty$, a size segregation takes place in such a way that the network results into a collection of homogeneous regions of different labels; however on each homogeneous region, sites and bonds have virtually the same numeric label. Here, it is considered that during tablet fabrication, the network self-organizes in such a way that drug grains of mean size l_0 are formed.

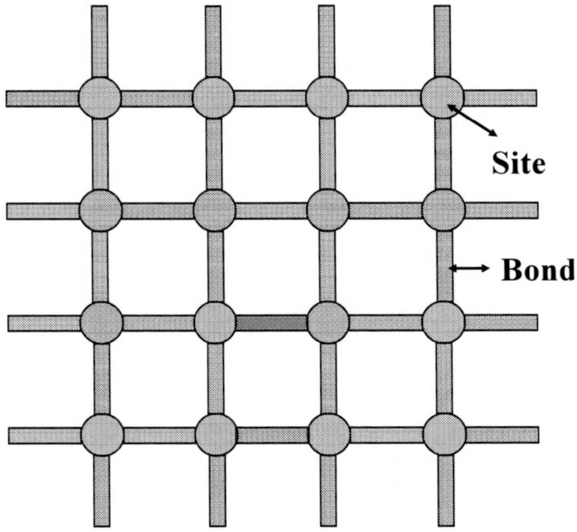

Figure 3. Schematic representation of the porous space in terms of sites and bonds. Here, a cubic lattice represents the porous medium where sites are located at nodes while each bond connection is located in between two adjacent sites.

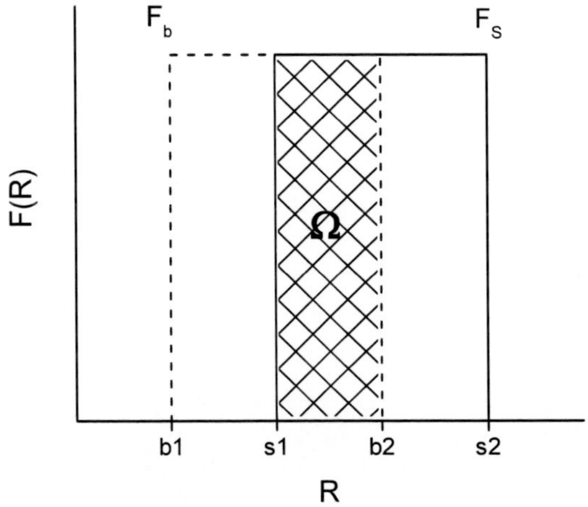

Figure 4. Uniform probability densities F_s and F_b as a function of R, for sites and bonds respectively, showing the overlap Ω between them.

Equation (3) is only valid for square networks, and no relationship between l_0 and Ω has been established for cubic networks. That is why, to look into the way by which network elements are interlinked inside a simulated three-dimensional Site-Bond lattice, a correlation length is calculated as expounded in [20]. For that, it is assumed that the numeric label correlation coefficient, $C^{xy}(u)$, for two net elements x and y separated by a distance u (in lattice units), is given by the following expression:

$$C^{xy}(u)=\frac{\langle (R_x-\bar{R}_x)(R_y-\bar{R}_y)\rangle}{\left[\langle (R_x-\bar{R}_x)^2\rangle\langle (R_y-\bar{R}_y)^2\rangle\right]^{1/2}} \tag{4}$$

where R_i represents an element numerical label, and x, y being either S (a site) or B (a bond). This correlation function can be measured through Monte Carlo simulation by using a network of a finite size L. In this work the spatial correlation of the system is characterized via the site-site correlation length, ξ_{SS}, which is calculated by means of the expression proposed in [22]:

$$C^{SS}(u)=\exp\left(-\frac{u}{\xi_{SS}}\right) \tag{5}$$

where $C^{SS}(u)$ is the numeric correlation coefficient between two sites separated by a distance u.

Next we describe how the initial conditions are established. As mentioned before, Ω makes possible to impose a spatial correlation among the sites forming the network. When a correlated network with the requested value of l_0 is constructed, the spatial distributions of D-particles and E-particles inside the network proceed as following, c.f. figure 5. The normalized area below F_s has two real limits, s_1 and s_2, i.e. the minimum and the supreme of the F_s set, respectively (c.f. figure 4):

$$\int_{s_1}^{s_2} F_s \, dR = 1 \tag{6}$$

There is an equivalent relationship among b_1, b_2 and F_b. Then, it is possible to match a portion from the F_s normalized area with a value equal to the desired drug concentration C_0:

$$C_0 = \int_{s_1}^{s_i} F_s \, dR \tag{7}$$

where $s_i \in [s_1, s_2]$. As, at this stage, all the sites of the network are labeled with numbers coming from F_s, then, as illustrated in figure 3, the allotment of drug and excipient particles to the sites of the network is achieved straightforwardly; those with an assigned number smaller or equal to a threshold value are considered D-sites and those with an assigned number larger than the threshold are considered E-sites. It is pertinent to mention that: (i) after correlating the D-site or E-site assignations, the number R given to each site does not have any more relevance during the simulation, (ii) D-particles are considered indistinguishable, and (iii) we interpret the l_0 value as the characteristic size of the drug grains in the device. Finally, the number of leaking sites, N_{leak}, the number of sites in the network, N_{total}, and the number of sites at the lattice border, N_{border}, are counted up and registered.

23.68	16.80	21.66	22.75	15.08	16.08
17.39	21.18	17.39	19.01	20.77	18.17
15.90	19.93	23.58	17.83	18.66	19.51
18.96	14.98	16.74	16.50	14.95	14.22
17.02	15.61	23.48	22.94	17.07	23.81

E	D	E	E	D	D
D	E	D	E	E	E
D	E	E	D	E	E
E	D	D	D	D	D
D	D	E	E	D	E

Figure 5. Schematic representation of the initial conditions in a cubic network. For instance, when $\Omega = 0.6$, the sites of the lattice are labeled with numbers in the interval $(14,24)$, in the numeric experiences these figures are float. To get a $C_0 = 0.4$, it is necessary to choose a threshold equal to 18, i.e. the sites labeled with a number smaller or equal to 18 are considered D-sites, else they are considered sites occupied by excipient, E-sites.

RESULTS AND DISCUSSION

In Table 1 the calculated d_w values, on the used Menger sponges, are presented. Their relative error, $r.e.$, is bounded by 0.005. In this way, we have achieved three-dimensional fractal structures with $d_f = 2.727$, which is a plausible drug dosing value. Additionally, d_w values cover the range $d_w \in (2.02,3.00)$. Another useful parameter is N_{leak}/N_{total}, where N_{leak} denotes the number of drug escaping sites and N_{total} is the total number of sites in the matrix platform. The M_t/M_0 (drug released fraction) values corresponding to the various Menger sponges are shown in figure 6 as a function of time. Note that porous structures, having the same fractal dimension but endowed with different topological properties, produce different release profiles. In addition, figure 6 shows fitted lines (according to equation (2)) for each data set.

Table 1. Drug release parameters corresponding to Menger sponges, $d_f = 2.727$. Delivery occurs through the six faces of the cubic matrix platform

Sponge	$\frac{M_\infty}{M_0}$	d_w	$\frac{N_{leak}}{N_{total}}$	a	b	d_s^*	d_s
ASM	0.962	2.455	0.107	0.056	0.548	0.938	2.222
BSM	0.985	2.194	0.132	0.064	0.575	0.878	2.486
CSM	1.000	2.028	0.066	0.020	0.759	0.398	2.689
DSM	0.971	2.366	0.133	0.071	0.552	0.974	2.305
ESM	0.896	2.761	0.092	0.048	0.540	0.985	1.975
FSM	0.866	2.998	0.083	0.041	0.572	0.979	1.819

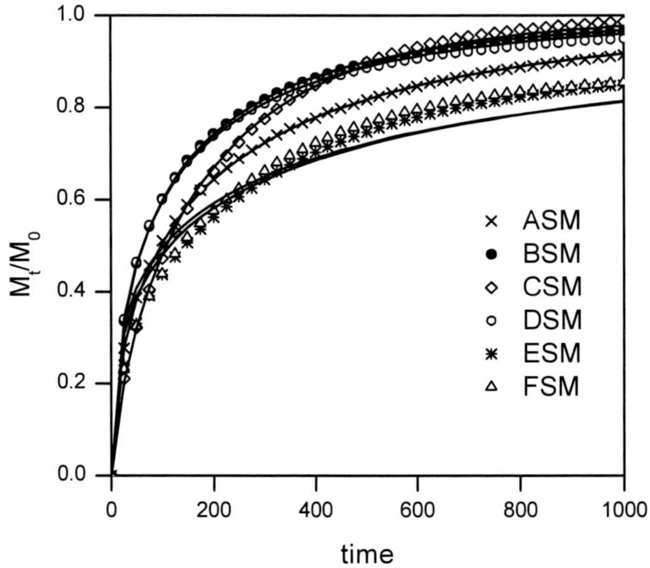

Figure 6. Drug release from Menger sponges. Symbols represent numerical results, while solid lines show the fitting of these data through the Weibull equation.

Table 1 also lists the estimated a, and b values. The relative errors are 0.02 for a, and 0.005 for b. In general, there is a very good accord between the plotted data and the predicted behavior advanced by equation (2). This means that drug release from finite Menger sponges can be well represented by the Weibull equation. This agrees with the observations of other authors [12, 13]. Also note the differences among the diverse M_∞ / M_0 values given in Table 1. Here, the maximum time in calculations is 10^5 Monte Carlo steps. The dissimilar amounts of drug entrapped inside the blind porosity (i.e. pores which are not connected to the exterior) of each sponge is responsible for these differences.

The a and b parameters are related to the structural properties of the matrix platform. As mentioned elsewhere [11], the values of a and N_{leak}/N_{total} obey a linear relationship. In our case, this expression has the following form:

$$a = -0.017 + 0.651\frac{N_{leak}}{N_{total}} \quad (8)$$

Following [11], notice that the independent (-0.017) is quite small with respect to the dependent term (0.651), and thus the value of a is mostly determined by the N_{leak}/N_{total} ratio; therefore we may conclude that $a \propto N_{leak}/N_{total}$. In contrast, our results show no linear relationship between b and N_{leak}/N_{total}. Then b values should have two contributions: (i) b should be proportional to the specific surface area of the matrix, since a high value for this parameter means that there are many exits for drug escape; and (ii) b should be a function of the ability of the drug particles to travel inside the matrix platform.

A drug release problem can be seen as a study of the kinetics of the reaction $A + B \rightarrow C$ [11], where A represents traveling particles while B and C are regarded as static ones; the above scheme portrays the well known trapping problem [15]. In the fractal approach there are three important dimensions: d_f, d_w, and the fracton dimension d_s. The last one takes into account the way by which a diffusing particle "sees" the heterogeneities present in the porous medium during its random walking transit, and is related to d_f and d_w by:

$$d_s = 2d_f/d_w \quad (9)$$

It is pertinent to note that there are two significant differences between the drug release problem and the classical trapping problem: (i) during drug release, the traps are not randomly distributed throughout the porous medium. Instead, they are mainly located at the device boundaries. In fact, the boundary fraction, which is part of the embedded drug clusters, constitutes the trap sites; and (ii) During trapping, the porosity of the system, ε, is not changing greatly, whereas in drug release the porosity of the tablet changes notably. It is then expected that $d_s > d_s^*$, where d_s^* is the *fracton dimension* of the drug release problem. d_s^* is an effective fracton dimension as those computed by Argyrakis and Kopelman [23]. In this work, following [15] the values of d_s^* are calculated from the slopes of the $\log((dQ/dt)/N(t))\, vs \log t$ curves plotted in figure 7. The obtained d_s^* values are presented in Table 1. Notice that the d_s^* values are very different to those computed from equation (9). As mentioned earlier, the reasons for this difference are the spatial segregation of the leaking sites and the dynamic behavior of porosity value. The fit of these data shows that b has a linear dependency with d_s^*.

$$b = 0.907 - 0.374 d_s^* \tag{10}$$

This equation clearly shows that b is truly related to the transport properties of the drug delivery system.

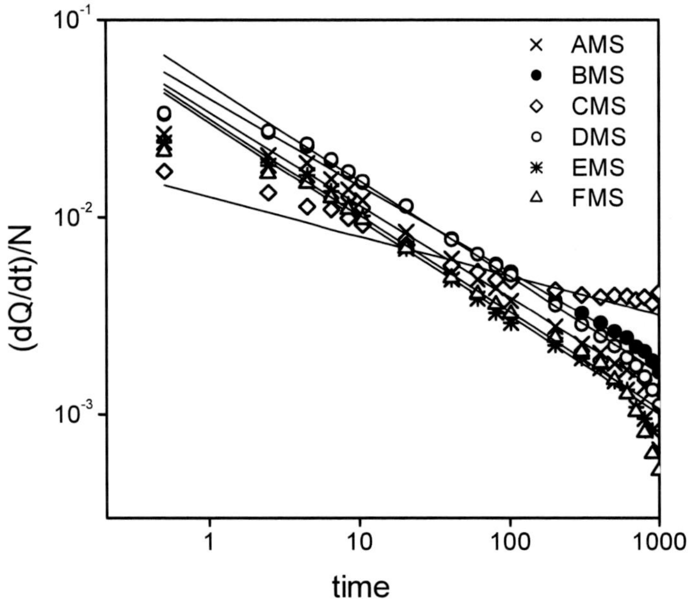

Figure 7. Determination of d_s^* values. Symbols represent numerical results, while solid lines show the fitting of these data.

Next the results coming from the study on the cubic networks are presented. The figure 8 shows simulation results of the rate of D-particles release as a function of time, in the case of release through only one face of the cubic lattice and from a matrix containing 20^3 sites with $C_0 = 0.5$ and $l_0 = 0.0$. In the same figure, we include the data obtained from [25] under similar thermodynamic conditions, but resting on a shorter time interval, $t \in [0, 2000]$. Time interval in Figure 8 is $t \in [0, 3000]$ in order to show details, however our simulation halts when more than 90% of the D-particles have been released from the matrix, i.e., about 20000 MCS. Notice the qualitative similarity between both sets of data, for $t < 200$ there is a very fast decrease of $\frac{dQ}{dt}$ value as t goes up, and when $t > 200$ this trend is gradually mitigated until a linear drop correlation between $\frac{dQ}{dt}$ and time is established, this last behavior agrees well with the observations done in [11]. However, note also the quantitative difference between these two data sets. Initially, $t < 50$, D-particles placed on the lattice border are released fast, hence this early step is mainly controlled by the value of the specific area, $\frac{N_{leak}}{N_{border}}$; here both models have very similar behaviors. When the soluble material placed near the lattice border is exhausted, drug must be transported from the interior of the release device, and so, gradually mass transport inside the porous matrix becomes the most relevant transport mechanism. In [9] the matrix porosity is practically invariable and all the matrix sites are accessible to D-particles, whereas in this work the matrix porosity starts from cero, and has a monotonous increase as release occurs, moreover only half of the network sites are potentially opened to D-particles, and the rest are totally closed to them. Thus, our simulations present a more difficult diffusion (sub-diffusive behavior) of D-particles inside the matrix core, and then a less efficient release than that occurring from an all-open network. In brief, within this time interval, $t \in [200, 2000]$, our simulation presents a lower reduction of $\frac{dQ}{dt}$ as t increases than the simulation results given in [9]. Finally, when $t > 2000$, the all open porous media is almost empty, the escape probability drops to zero, and a crossing between the two sets of data is produced, i.e., for $t > 2000$, the $\frac{dQ}{dt}$ value from our simulation is higher than the akin value given by [9].

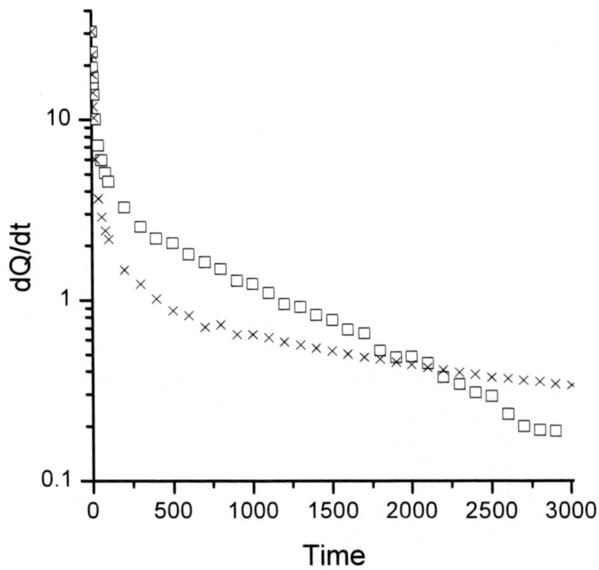

Figure 8. Plot of the release rate dQ/dt vs time. The lattice size is 20^3 and the initial concentration of drug particles is C_0=0.5. Dotted line is the result given by Bunde *et al.* [25], while the solid line is the result of the current simulation averaged over 10^3 simulation experiments.

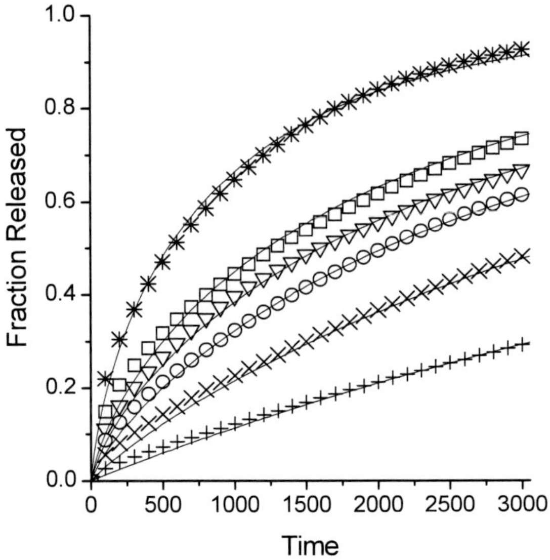

Figure 9. Release profiles from two opposite faces of cubic matrices, at various L values, and $C_0 = 0.594$. The symbols represent the Monte Carlo simulation data, while solid lines are the corresponding fitting by Weibull model. $L = 27$ (□), $L = 40$ (□), $L = 45$ (∇), $L = 50$ (o), $L = 60$ (×), and $L = 81$ (+).

Table 2. Simple cubic networks. Release through two opposite faces. The values of N_{leak} correspond to the exposed area, i.e., two faces. $C_0 = 0.594$. a and b are the amounts defined in Equation 2. The relative error of a and b are bounded by 0.05 and 0.02, respectively

L	$\frac{N_{leak}}{N_{total}}$	a	b
27	0.044	0.0102	0.678
40	0.030	0.0035	0.743
45	0.026	0.0026	0.757
50	0.024	0.0015	0.805
60	0.020	0.0005	0.895
81	0.015	0.0002	0.967

Figure 9 shows the effect of L on the release profile occurring through two opposite faces of the lattice platform, and when $C_0 = 0.594$. In agreement with [12], figure 9 shows that the release profile is a function of L. In all the cases shown here, there is first a time interval showing fast release kinetics, second there is a transition zone where the release rate goes down, and finally there is a zone characterized by slow release kinetics. This behavior can be explained as follows. During the release process, drug should be first dissolved and after that transported in the liquid phase. At first, the drug placed at the device's border is quickly transferred into the liquid phase and released to the surroundings, so this step is characterized by a quick release and mainly controlled by the specific surface area of the device. Nevertheless, when this material is exhausted, drug has to be transported from the core of the device; hence the release is more and more controlled by mass diffusion. This gradual change on the control mechanism produces both, first the transition zone, and second the slow release zone wholly controlled by drug diffusion. In brief, the combined effects of these two mechanisms determine the release profile. These observations are confirmed by data in Table 2, where N_{leak}/N_{total} values are given by the release through two opposite faces of simple cubic networks for $L \in \{27, 40, 45, 50, 60, 81\}$ in lattice units, with $C_0 = 0.594$. Notice that $\frac{N_{leak}}{N_{total}} \rightarrow 0$ when $L \rightarrow \infty$, i.e., the portion of drug initially placed close to the network boundary decreases as L increases, so the meaning of the release controlled by drug mass diffusion increases as L increases, then the slow kinetics step takes place relatively earlier as L increases. Note also that in all the presented cases, it is possible to achieve a quite accurate fitting of the simulation results using Equation 2. Again, Weibull function is a satisfactory option to model controlled release. Table 2 gives the estimated values of a, and b. The relative errors of a and b are bounded by 0.05 and 0.02, respectively. It turns out that: (i) the a factor takes values from 0.0070 to 0.0002 as L goes up from 27 to 81, i.e., the numbers in this table follow the same behavior as $\frac{N_{leak}}{N_{total}}$, and it mainly represents the influence of the device's specific surface area on the release [11,12], and (ii) the stretching exponent b takes values in the range 0.678 to 0.967, i.e. larger values than those given in [11,12] for systems with higher dimensionality, and then easier mass transfer. It is pertinent to bring up that in reference [24] the authors explicitly study release from coated theophylline particles and present results from an experimental drug release study for different values of coating and plasticizer added to the coating polymer. The values of b exponent are in all cases in the range of 0.54 to 1.18, depending on the amount of coating and plasticizer, notice that all the results of our simulations fall inside this last range.

On the other hand, it has been already mentioned, the power law model, Equation 1, is believed to describe accurately the release at short times. Figure 10 shows our simulation results and fittings with the Weibull function and the power law model. As signaled in [11], the Weibull equation describes quite well the entire release process, whereas the power law diverges after a certain point in time, and its utility is limited by the initial part of the release curve.

Figure 11 shows the drug release profiles obtained from systems characterized by various correlation lengths, $l_0 \in \{0.17, 0.68, 1.07, 1.35, 1.88, 2.99, 4.01\}$, in lattice units, for L=27. Remember that the higher is the correlation length, the higher is the mean extent of the original drug grains (drug clusters). In general, the higher the l_0 value is, the faster the release is, because a higher l_0 value means stronger interactions among D-particles, so the release of a D-particle promotes the transfer of other D-particle, making their release easier as l_0 grows.

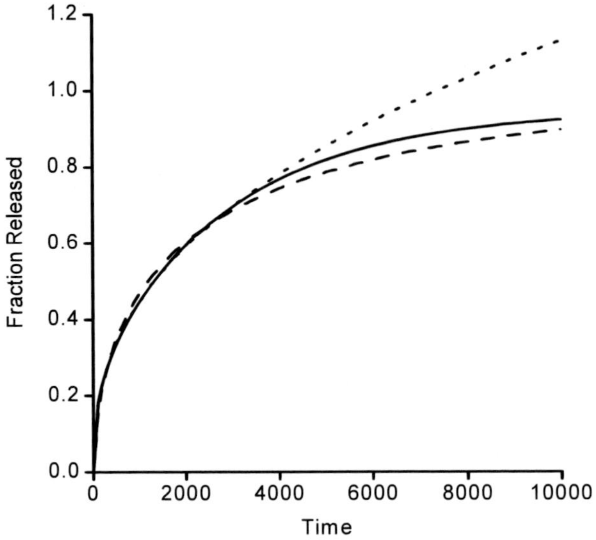

Figure 10. Plot of cumulative amounts of drug released from cubic matrices with L=27 and through two opposite faces as a function of time. Solid line, Monte Carlo simulation; dotted line, power law fitting; dashed line, Weibull model fitting.

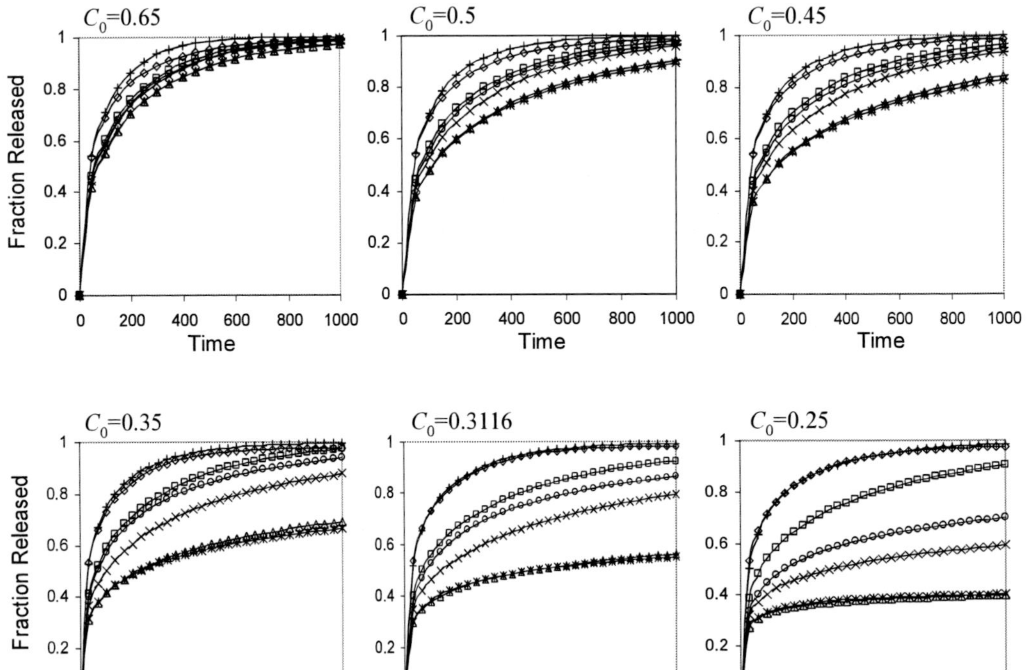

Figure 11. Results from Monte Carlo simulation. Fraction of dose released from cubic matrices by their total surface area with diverse initial drug load (C_0) and several correlation lengths, l_0. (+) $l_0 = 4.01$, (◇) $l_0 = 2.99$, (□) $l_0 = 1.88$, (o) $l_0 = 1.35$, (x) $l_0 = 1.07$, (Δ) $l_0 = 0.68$, (*) $l_0 = 0.17$.

Furthermore, when C_0 goes down, the effect of l_0 is more significant. First, when $C_0 = 0.65$, practically all D-particles in all the networks are connected to the environment, i.e., there is no drug trapping. However, when C_0 value decreases the dose fraction that is trapped increases. This effect is more evident as the value of l_0 is smaller because the percolation threshold is a function of l_0 [19, 25], and decreases as l_0 value increases; in fact, the percolation threshold was roughly calculated as 0.251, 0.244, 0.146, 0.083, 0.016, 0.012 and 0.007 [4] for l_0 values equal to 0.17, 0.68, 1.07, 1.35, 1.88, 2.99, and 4.01, respectively. Observe that the profile corresponding to $l_0 = 4.01$ and $l_0 = 2.99$, the highest l_0 values, are little affected by the C_0 value; on the contrary, the other drug outlines change a lot.

As a whole, there is a very good agreement between our Monte Carlo simulation data and Equation (2). The parameters a and b of this equation are somehow connected to the properties of the matrix platform. Figure 12 presents the obtained a values as a function of the $\frac{N_{leak}}{N_{total}}$ values, and also the best fitted straight line for each l_0 value, the equations of these fitted lines are given in Table 3, they are computed by considering only the figures corresponding to C_0 values higher than the respective percolation threshold. For drug concentration values smaller than the network percolation threshold, the behavior is very different, i.e. there is not a linear relationship between a and $\frac{N_{leak}}{N_{total}}$. In reference [12] a values and $\frac{N_{leak}}{N_{total}}$ values get a positive linear relationship, whilst our results have negative linear relationships. This difference could be explained as follows, in [12] the used matrices are Euclidean (all the sites are accessible) while our drug-excipient systems are *LDS*. In Euclidean networks, all the release sites have the same relevance, while in a *LDS*, the significance of a release site is a function of the mean extent of the cluster where it belongs, hence if C_0 grows, the number of important release sites does not grow in the same proportion than $\frac{N_{leak}}{N_{total}}$ value, producing a linear and negative relationship between a value and $\frac{N_{leak}}{N_{total}}$ value. Notice that, in general, the slope of this trend goes down as l_0 grows, for instance when $l_0 = 0.17$ we find $a = 0.094 - 0.311\frac{N_{leak}}{N_{total}}$, and when $l_0 = 4.01$ we find $a = 0.129 - 0.542\frac{N_{leak}}{N_{total}}$. Notice also that for the same C_0 value, a decreases as l_0 increases. In brief, we may conclude that in *LDS*, the parameter a is determined by both, the specific surface area of the device and its l_0 value. Finally, figure 13 shows, for each l_0

value, a linear and positive relationship between b values and $\frac{N_{leak}}{N_{total}}$ values, the equations of these relationships are given in Table 2, as before the lines are computed considering only the data that correspond to a C_0 value higher than the respective drug percolation threshold. Under this threshold, there is not a well established tendency between b and $\frac{N_{leak}}{N_{total}}$. From data in figure 13, we find that, for a given l_0 value, b seems to be essentially a function of $\frac{N_{leak}}{N_{total}}$ value. For instance, we find $b = 0.207 + 2.722\frac{N_{leak}}{N_{total}}$ and $b = 0.385 + 2.146\frac{N_{leak}}{N_{total}}$, for $l_0 = 0.17$ and $l_0 = 4.01$, respectively. Observe that in these equations, the slope is larger than the independent term. This tendency is more evident as the value of l_0 decreases. Indeed, the value of b should include two contributions: (i) it should be proportional to the specific surface area since a high specific surface means that there are many exits, so it is easier to find an escaping route, and (ii) it should be a function of the ability of the drug particles to move inside the matrix platform. This last point is confirmed by the effect of the correlation length over b (see figure 13), i.e., at the same C_0 value, the higher l_0 value is, the higher the b value is, because a higher l_0 means an easier mass transfer inside the matrix. Also notice that the slope of the fitted line goes down as l_0 grows. Therefore we may conclude that b is also determined by both the specific surface area and the internal topology of the matrix, i.e., the l_0 value.

Table 3. The equations of lines fitted to data presented in figure 8 and figure 9, for various l_0 values

l_0	$a = a\left(\frac{N_{leak}}{N_{total}}\right)$	$b = b\left(\frac{N_{leak}}{N_{total}}\right)$
0.17	$a = 0.094 - 0.311\frac{N_{leak}}{N_{total}}$	$b = 0.207 + 2.773\frac{N_{leak}}{N_{total}}$
l_0	$a = a\left(\frac{N_{leak}}{N_{total}}\right)$	$b = b\left(\frac{N_{leak}}{N_{total}}\right)$
0.68	$a = 0.078 - 0.210\frac{N_{leak}}{N_{total}}$	$b = 0.226 + 2.430\frac{N_{leak}}{N_{total}}$

1.07	$a = 0.064 - 0.091 \frac{N_{leak}}{N_{total}}$	$b = 0.401 + 1.528 \frac{N_{leak}}{N_{total}}$
1.35	$a = 0.088 - 0.252 \frac{N_{leak}}{N_{total}}$	$b = 0.392 + 1.619 \frac{N_{leak}}{N_{total}}$
1.88	$a = 0.089 - 0.272 \frac{N_{leak}}{N_{total}}$	$b = 0.398 + 1.640 \frac{N_{leak}}{N_{total}}$
2.99	$a = 0.143 - 0.470 \frac{N_{leak}}{N_{total}}$	$b = 0.379 + 1.948 \frac{N_{leak}}{N_{total}}$
4.01	$a = 0.129 - 0.542 \frac{N_{leak}}{N_{total}}$	$b = 0.385 + 2.146 \frac{N_{leak}}{N_{total}}$

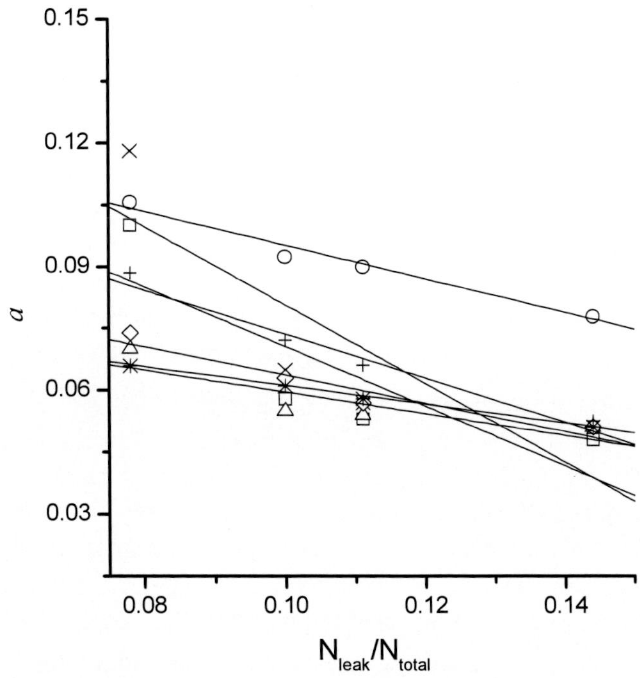

Figure 12. Parameter a vs N_{leak} / N_{total} for several correlation lengths. These results were obtained by exposing the total surface area of the device.

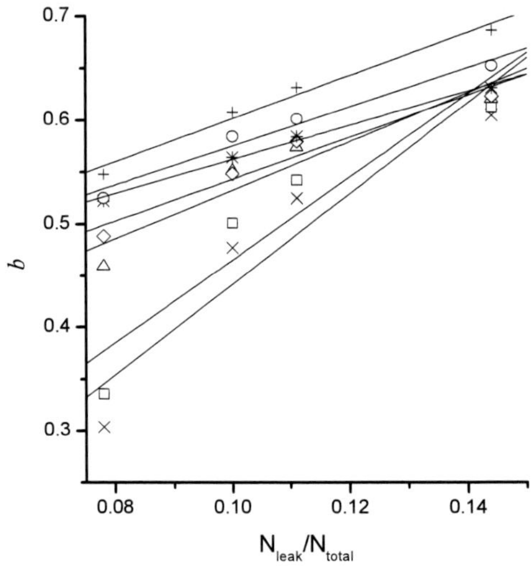

Figure 13. Parameter *b* vs. N_{leak} / N_{total} for different correlation lengths. These results were obtained by exposing the total surface area of the device.

Conclusions

Overall, this study provides virtual evidence for the successful use of the Weibull function in drug release studies. Drug release from Menger sponges is characterized by a non-Fickian behavior. Nevertheless, this abnormal process can be well described in terms of a Weibull equation, in which the device surface is defined by a value, while the device transport properties are essentially defined by b value. From the results of the simulation of drug release from cubic networks, it was found that the drug-excipient ratio is a factor that determines the release mechanism from a matrix system. A sub-diffusive behavior of the drug inside the matrix was due to the presence of the excipient. Another factor that modified the release profile was the N_{leak}/N_{total} value. It was found that the N_{leak}/N_{total} value is directly related to the surface/volume ratio of a matrix device. Since this last value is a function of the matrix size, the size of the matrix system affected the drug release profile too.

On the other hand, the DSBM is a simple model (in fact the simplest, to our knowledge) capable of describing random media with different topological structures. These are generated by varying a single parameter, Ω, the overlapping between the site and bond probability densities, while the details of the porous medium will depend, of course, on the shape of these distributions. The parameter Ω can be associated to a correlation length l_0, in such a way that $l_0 \to 0$ as $\Omega \to 0$ and $l_0 \to \infty$ as $\Omega \to 1$. From this concepts, it has been possible to simulate drug release from porous networks topologically equivalent to a granular structure, where l_0 was the mean granule size. Again, we found an excellent fitting between our drug release results and the Weibull function. The values of parameters *a* and *b* of this equation were strongly dependent on: (i) the specific surface area, and (ii) the internal topology of the matrix. How does topology affect the percolation properties of the medium? In a similar way as for the Cayley tree, in this work, we found that percolation probabilities increase and percolation thresholds decrease as Ω increases. In this way, it was illustrated that the clarity of the concepts involved in the DSBM could be easily applied or associated with percolation theory, which helps us to understand solid matrix systems as drug controlled release platforms. Finally, the critical modeling of drug release from matrix-type delivery systems is important in order to understand the implicated transport mechanisms, and to predict the effect of the device design parameters on the release rate.

NOMENCLATURE

a	Dimensionless real number
b	Dimensionless real number
b_i	Element of F_b
C_0	Initial drug concentration, fraction of sites occupied by drug
$C^{xy}(u)$	Correlation coefficient
D	Drug
DSBM	Dual site-bond model
d	Euclidean dimension
d_f	Fractal dimension
d_s	Fracton dimension of the trapping problem
d_s^*	Fracton dimension of the release problem
d_w	Random walk fractal dimension
dQ/dt	Release rate
E	Excipient
ε_t	Matrix porosity at time t
F_s	Uniform real numbers distribution for sites
F_b	Uniform real numbers distribution for bonds
L	Characteristic lattice length, lattice units
LDS	Low dimensional system
l_0	Correlation length, lattice units
MCS	Monte Carlo step
M	Dimensionless mass
M_∞	Dimensionless mass of drug released at time infinity
M_t	Dimensionless mass of drug released at time t
N_{border}	Number of sites on the network boundary
N_{leak}	Number of trapping sites
N_{total}	Number of sites in the network

N_t	Number of drug particles remaining in the cubic lattice at time t
R	Dimensionless real number
R_i	Numeric label of the element i
s_i	Element of F_s
t	Dimensionless time
u	Distance in lattice units
Ω	Overlapping between F_b and F_s

REFERENCES

[1] K. Takada, and H. Yoshikawa. Oral Drug Delivery, Traditional. In Encyclopedia of Controlled Drug Delivery, 1st Ed.; Mathiowitz E. (ed), John Wiley & Sons, Inc. New York, 728-742, 1999.

[2] H. Leuenberger, R. Leu, and J. D. Bonny. Application of percolation theory and fractal geometry to tablet compaction. *Drug Dev. Ind. Phar.* 18:723-766, 1992.

[3] H. Leuenberger, J. D. Bonny, and M. Kolb. Percolation effects in matrix-type controlled release systems. *Int. J. Pharm.* 115:217-224, 1995.

[4] J. D. Bonny, and H. Leuenberger. Matrix type controlled release systems: II. Percolation effects in non-swellable matrices. *Pharm. Acta Helv*. 68:25-33, 1993.

[5] A. Bunde and S. Havlin, Percolation I, in Fractals and Disordered Systems, A. Bunde and S. Havlin Eds. 2° Edition. Springer-Verlag, New York, 1996, Chapter 2.

[6] R. Villalobos, A. Ganem, S, Cordero, A.M. Vidales, A. Domínguez, Effect of drug-Excipient Ratio in Matrix-Type-Controlled Release Systems: Computer Simulation Study. *Drug Dev. Ind. Phar*, 31:535-543, 2005.

[7] R. Villalobos, A.M. Vidales, S. Cordero, D. Quintanar, A. Domínguez, Monte Carlo Simulation of Diffusion-limited Drug Release from Fractal Matrices, *J. Sol-Gel Sci. Techn*, 37:195-199, 2006.

[8] R. Villalobos, S. Cordero, A.M. Vidales, A. Domínguez, In Silico Study on the Effects of Matriz Structure in Controlled Drug Release, *Physica A*, 367: 305-318, 2006.

[9] A. Bunde, S. Havlin, R. Nossal, H. E. Stanley, and G. H. Weiss. On controlled diffusion-limited drug release from a leaky matrix. *J. Chem. Phys*. 83:5905-5913, 1985.

[10] P. Costa, and J. M. Sousa. Modeling and comparison of dissolution profiles. *Eur. J. Pharm. Sci.* 13:123-133, 2001.

[11] K. Kosmidis, P. Argyrakis, and P. Macheras. Fractal kinetics in drug release from finite fractal matrices. *J. Chem. Phys.* 119:6373-6377, 2003.

[12] K. Kosmidis, P. Argyrakis, and P. Macheras. A reappraisal of drug release laws using Monte Carlo simulations: the prevalence of the Weibull function. *Pharm. Res*. 20:988-995, 2003.

[13] V. Papadopoulou, K. Kosmidis, M. Vlachou, P. Macheras, On the Use of the Weibull Function for the Discernment of Drug Release Mechanisms, *Int. J. Pharmaceutics*, 309:44-59, 2006.

[14] S. Havlin. Molecular diffusion and reactions. In Avnir D (ed), The Fractal Approach of Heterogeneous Chemistry, John Wiley & Sons, Chichester, 251-269, 1989.

[15] R. Kopelman, Diffusion-controlled reaction kinetics. In Avnir D (ed), The Fractal Approach of Heterogeneous Chemistry, John Wiley & Sons, Chichester, 295-309, 1989.

[16] M. Usteri, J.D. Bonny, H. Leuenberger, Pharm. Acta Helv., 65: 55-61, 1993.

[17] S. Havlin, D. Ben-Avraham, Diffusion in Disordered Media. Adv. Phys., 36: 695-798, 1987.

[18] S. Tarafdar, A. Franz, Ch. Schulzky, and K.H. Hoffmann. Modelling porous structures by repeated Sierpinski carpets. Physica A., 292: 1-8, 2001.

[19] I. Kornhauser, F. Rojas, R. J. Faccio, J. L. Riccardo, A. M. Vidales, and G. Zgrablich. Structure characterization of disordered porous media – a memorial review dedicated to Vicente Mayagoitia. *Fractals*. 5:355-377, 1997.

[20] S. Cordero, F. Rojas, and J. L. Riccardo. Simulation of three-dimensional porous networks. *Colloids Surf. A.* 187-188:425-438, 2001.

[21] R.H. López, A.M. Vidales, G. Zgrablich. Correlated-Site-Bond Ensembles: Estatistical Equilibrium and Finite Size Effects. *Langmuir*. 16: 3441-3445, 2000.

[22] J. L. Riccardo, V. Pereyra, G. Zgrablich, F. Rojas, V. Mayagoitia, I. Kornhasuser. Characterization of Energetic Surface Heterogeneity by Dual Site-Bond Model. *Langmuir*. 9:2730-2736, 1993.

[23] P. Argyrakis, R. Kopelman, Fractal to Euclidean Crossover and Scaling for Random Walkers on Percolation Clusters, *J. Chem. Phys.* 81:1015-1018, 1984.

[24] C. Felipe, R.H. lopez, A.M. Vidales, and A. Dominguez. 2D Automaton simulation of bubble growth by solute diffusion, *Adsorption,* 11: 491-496, 2005.

[25] I. Antal, R. Zelco, N. Roczey, J. Plachy and I. Racs. Dissolution and Diffuse Reflectance Characteristics of Coated Theophylline Particles. *Int. J. Pharmacol.* 155:83-89 (1997).

In: Porous Media: Heat and Mass Transfer... ISBN: 978-1-60692-437-2
Editors: J. L. Acosta and A.F. Camacho

Chapter 5

MODELING OF SURFACTANT AND CONTAMINANT TRANSPORT IN POROUS MEDIA

***Hefa Cheng*[1], *Jian Luo*[2], *and Gang Chen*[*3]**

[1] Department of Civil and Environmental Engineering
Stanford University; Stanford, CA, U.S.A. 94305
[2] School of Civil and Environmental Engineering
Georgia Institute of Technology; Atlanta, GA, U.S.A., 30332
[3] Department of Civil and Environmental Engineering
FAMU-FSU College of Engineering
Tallahassee, FL, U.S.A. 32310

ABSTRACT

Surfactant-enhanced aquifer remediation (SEAR) is an excellent source-depletion technology for the remediation of the subsurface contaminated by non-aqueous phase liquid (NAPL). Modeling of surfactant and contaminant transport in the porous media can provide quantitative assessment and prediction of SEAR performance. Processes including surfactant micellization, surfactant and contaminant sorption on the aquifer media, contaminant micellar-solubilization, contaminant adsolubilization by admicelles/hemimicelles formed on the aquifer media, advective and dispersive transport of the surfactant and contaminant, reduction of contaminant residual saturation, and alteration of interfacial properties may be involved in SEAR. The reaction kinetics and heterogeneities of the aquifer properties further complicate modeling of these processes. A number of models have been developed to describe the transport behavior of surfactant and contaminant, which include surfactant and contaminant coupled transport model, solute transport model coupled with two- or multi-site sorption, multi-component reactive transport model, multi-component multi-phase flow model, and UTCHEM model, etc. In this review, assumptions and reactive processes as well as the mathematical formulations of these models are reviewed. It is recommended that the processes involved in SEAR be characterized independently with controlled experiments to examine their individual contribution to the overall contaminant fate and transport. Increased understanding of the fundamental

* Corresponding author phone: (+1) 850-410-6303; e-mail: gchen@eng.fsu.edu

hydrodynamic and physiochemical processes occurring during SEAR will significantly improve the accuracy of SEAR modeling.

1. INTRODUCTION

The wide spreading of contamination in groundwater aquifers by non-aqueous liquid (NAPLs) has led to intensive studies on the transport and fate processes of these contaminants in the subsurface environment and the potential remediation protocols [5, 18, 21, 28, 30]. NAPLs frequently enter the vadose zone as discrete liquid phases that migrate downwards as a result of gravitational forces and capillary forces. When a large amount of organic liquid is spilled, it may eventually penetrate the capillary entry barrier in the vadose zone and reach the groundwater table [37]. NAPLs with density greater than water (DNAPLs) will continue to migrate vertically through the saturated zone until being stopped by impermeable aquitards [36]. On the other hand, NAPLs that are lighter than water (LNAPLs) spread laterally along the water table, forming floating pools of organic phase due to buoyancy effect. Flutuations of the groundwater table results in the vertical re-distribution of NAPLs and formation of "smear zones" in the vadose and saturated zone [19, 20, 25].

As the NAPLs migrate in the subsurface, a portion of the organic liquid will be trapped within the soil pores as immobile ganglia or blobs by the capillary forces, which are influenced by pore geometry, interfacial tension (IFT) and media surface wettability [2]. The residual organic phase in the soil pores poses a long-term source of groundwater contamination. Non-uniform NAPL distribution, irregular groundwater flow patters, and rate-limited mass transfer between organic phase and groundwater in the heterogeneous aquifer limit the dissolution removal of the NAPLs by groundwater flow [31]. Traditional water flooding displace NAPLs trapped in large radius pores, but cannot remove the NAPL ganglia trapped in small pores where viscous forces of the driving water could not completely overcome the capillary forces holding the NAPLs in soil pores [4]. The ratio of viscous forces to capillary forces is termed as capillary number, which has been found to be correlated well with residual saturation. By raising the capillary number, the residual saturation of NAPLs can be greatly reduced. While traditional water flooding has been proved to be ineffective and expensive at cleaning up NAPL contamination, studies in various scales over the past decade indicate that SEAR can substantially enhance the water flooding processes and is an excellent alternative [13, 26, 29].

Surfactant solutions can increase the apparent solubility of the NAPLs by several orders of magnitude via micellar-solubilization, removing the residual NAPLs by enhanced dissolution (enhanced solubilization mechanism) [7, 11, 22]. Surfactant systems can also be designed to reduce the interfacial tension (IFT) between the NAPLs and water by orders of magnitude, thereby overcoming the capillary forces trapping the residual NAPLs and release them as both middle-phase and free phase (enhanced mobilization mechanis) microemulsion [3, 34, 38]. In field applications, either the enhanced solubilization or the enhanced mobilization mechanism is chosen depending upon the nature of the NAPLs on the site. Generally, the solubilization mechanism is preferred for the DNAPLs because of the concerns of vertical migration, while the mobilization mechanism via middle-phase microemulsion is most efficient for the removal of LNAPLs.

Organic contaminants dissolved in the aqueous solution are easily adsorbed on aquifer media during their migration in the subsurface. This adsorption is affected by the hydrophobicity of the contaminant, the organic content of the media, and the microporosity of the media as well [8-10, 14, 24]. While extra-micellar contaminants can be adsorbed directly, micellar-solubilized contaminants must partition out of micelles before being adsorbed. The interfacial activity and amphiphilic nature of the surfactant molecules also render them adsorbable to the aquifer media [17]. In general, anionic surfactants are preferred in SEAR to minimize surfactant sorption on soil and aquifer media [12, 16, 39].

Surfactant micellization, adsorption on aquifer media and formation of admicelles/hemimicelles, contaminant micellar-solubilization, middle phase micromulsion formation, contaminant adsorption on aquifer media, and contaminant adsorption on admicelles/hemimicelles (adsolubilization) will possibly occur during SEAR [17]. With the dissolution of residual contaminants and transport of surfactant solution and microemulsion phase, NAPL residual saturation and interfacial tension will decrease, resulting in contamination mobilization. Dissolution of NAPLs in the soil pores enhances the solution permeability of the local aquifer and the preferential flow paths for the surfactant solution. Non-equilibrium mass transfer and reaction kinetics during SEAR further complicate these processes. However, with increased understanding of the hydrodynamic and physicochemical processes involved in SEAR, mathematical modeling of surfactant and contaminant transport can provide quantitative assessment and prediction for the engineered systems [35].

2. Summary of Reviewed Models

2.1. Surfactant and Contaminant Coupled Transport Model

Danzer and Grathwohl [32] used a simple reactive transport model to simulate transport of polycyclic aromatic hydrocarbons (PAH) and surfactant in a laboratory column packed with natural aquifer material. They suggested that a linear isotherm below the critical micelle concentration (CMC) followed by a maximum sorption, q_{max}, described surfactant adsorption more closely, although a Langmuir type adsorption model was used in most literature. When surfactant concentrations were higher than the CMC, the adsorption of surfactants was described by:

$$q_{max} = K_{d,surf} \cdot CMC \tag{1}$$

where $K_{d,surf}$ is the linear partition coefficient of the surfactant. The sorption of contaminants on the aquifer media during transport was considered to be controlled by contaminant partitioning between the micelles and the aqueous phase (K_{mic}) and between the sorbed surfactants (hemimicelles and admicelles) and the aqueous phase (K_{adm}). In addition, it was also controlled by an apparent, time-dependent distribution coefficient ($K_{d,\ app}$) under non-equilibrium conditions. The retardation factor (R_d) was calculated as:

$$R_d = 1 + \frac{\rho_d}{n} \cdot \frac{K_{d,app} + K_{adm} \cdot q}{1 + K_{mic} \cdot C_{mic}} \tag{2}$$

where ρ_b and n are the bulk density and effective porosity of the media; q is the adsorbed surfactant concentration; and C_{mir} is the concentration of micelles.

In the column experiments, contaminants were pre-dissolved in the surfactant solution and the transport was found to be well represented by the advection-dispersion equation based on a local equilibrium assumption. It was also found that stepwise increase of the surfactant concentration above CMC was not retarded in the column, which agreed with the proposed surfactant adsorption isotherm shape. PAH apparent distribution ($K_{d,\ app}$) was measured under equilibrium conditions in batch systems. For the column experiments, it was found that the retardation of contaminant increased with increasing surfactant concentration up to the CMC, but it decreased with surfactant concentration once the CMC was exceeded.

2.2. Solute Transport Coupled with Two, Multi-Site Sorption Model

Smith et al. [33] employed a two-site sorption model to simulate their batch laboratory surfactant (Triton X-100) adsorption on a field soil. The model assumes that the adsorption behavior of surfactant follows Langmuir isotherm and that the sorption sites on the soil can be divided into equilibrium sites and kinetic sites. The governing equations for the equilibrium sites and kinetic sites were as follows:

$$\frac{\partial S_e}{\partial t} = F \frac{ab}{(1 + bC)^2} \frac{\partial C}{\partial t} \tag{3}$$

$$\frac{\partial S_k}{\partial t} = k[(1 - F)\frac{abC}{1 + bC} - S_k] \tag{4}$$

$$S = S_e + S_k = \frac{abC}{1 + bC} \tag{5}$$

where S_e and S_k are the sorbed surfactant concentrations for the equilibrium sites and kinetic sites, respectively; S is the total sorbed surfactant concentration; C is the surfactant concentration in the aqueous phase; F is the fraction of equilibrium sites; a and b are Langmuir parameters determined from equilibrium batch sorption experiments; k is the soil-water mass transfer coefficient; and t is the time. The kinetic sorption parameters F and k were determined by fitting the model with batch kinetic sorption data. During simulation of the laboratory column and field experiment data under steady state flow conditions, the following governing equation was used:

$$[1+F\frac{\rho_b ab}{(1+bC)^2}]\frac{\partial C}{\partial t}+\frac{\rho_b}{n}\frac{\partial S_k}{\partial t}=\nabla\cdot(D\nabla C)-\nabla\cdot(vC)+G \quad (6)$$

where D is the dispersion tensor; v is the average groundwater velocity vector; and G is an external supply. The initial and boundary conditions used for simulation of the column experimental data were as follows:

$$C(x,0)=S_e(x,0)=S_k(x,0)=0 \quad (7)$$

$$(c-\frac{D}{v}\frac{\partial C}{\partial x})\bigg|_{x=0}=\mathrm{C_{in}} \quad (8)$$

$$\frac{\partial \mathrm{C}(\infty,\mathrm{t})}{\partial \mathrm{t}}=\mathit{finite} \quad (9)$$

where C_{in} is the inflow surfactant concentration.

By calibrating the model parameters (D and v) to a non-reactive inorganic tracer, Smith et al. [33] found that adsorption of surfactants in the columns were rate-limited and the above model fitted the experimental data very well. Similar approaches were taken to simulate the field tests and reasonable agreements were observed. The authors also discussed the potential implications of non-equilibrium surfactant adsorption in SEAR, however, they did admit that the surfactants used (Triton X-100) in their research sorbed strongly to the field soil compared to anionic surfactants and other nonionic surfactants.

Sahoo et al. [27] investigated the rate-limited desorption of trichloroethene (TCE) from aquifer sediments and the effect of Triton X-100 on desorption and transport of TCE. Two sorption models (two-, and multi-site sorption) were applied in simulating the field TCE transport results. The governing equations incorporating the two-site sorption model were:

$$[1+\frac{F\rho_b K_d}{n}]\frac{\partial C}{\partial t}+\frac{\rho_b}{n}\frac{\partial S_k}{\partial t}=\nabla\cdot(D\nabla C)-\nabla\cdot(vC) \quad (10)$$

$$\frac{\partial \mathrm{S_k}}{\partial \mathrm{t}}=k[(1-F)K_d C-S_k] \quad (11)$$

The above governing equations subject to a zero-concentration gradient at all boundaries and to the following initial conditions:

$$C(x, y, 0) = C_i \tag{12}$$

$$\mathrm{S_k}(x, y, 0) = S_i \tag{13}$$

where F is the fraction of equilibrium sites; C is the aqueous TCE concentration; C_i is the initial aqueous TCE concentration; S_i is the initial sorbed TCE concentration; and S_k is the TCE concentration in the solid contributed by TCE sorbed on the kinetic sites (with a single-valued rate constant k).

A γ-probability distribution defined by a mean rate constant, k and a coefficient of variation, CV, was used to represent the distribution of kinetic rate constant for the multi-site model. NK discrete sites were used to represent the continuous site distribution with each site occupying 1/NK fraction of the soil. The governing equations incorporating the multi-site sorption model were:

$$\frac{\partial C}{\partial t} + \frac{\rho_b}{n}\frac{\partial S}{\partial t} = \nabla \cdot (D \nabla C) - \nabla \cdot (vC) \tag{14}$$

$$\frac{\partial \mathrm{S}}{\partial \mathrm{t}} = \sum_{k=1}^{NK} k_k [K_d C - S_k] \tag{15}$$

for the kth sorption site, the initial conditions are:

$$C(x, y, 0) = C_i \tag{16}$$

$$\mathrm{S_k}(x, y, 0) = \frac{S_i}{NK} \tag{17}$$

where S_K is the TCE concentration in the soil due to all the kinetic sites. The representative mass transfer coefficient, k_K, for each of the sites was obtained from the γ-probability function.

Uniform hydraulic conductivity, spatially variable dispersivity, homogeneous steady state flow, and transient solute transport were assumed in TCE transport simulation. The flow was assumed to be 2-D in the field site, with constant head boundaries at the upper and lower boundaries and no flow boundaries on the side boundaries far away form the injection wells. The flow was essentially horizontal during the injection and pumping. For TCE transport, a zero-concentration gradient boundary condition was assumed at all boundaries, which were away from the region of interest. TCE transport was not coupled to that of surfactant, instead, contaminant sorption parameters were changed abruptly once surfactant is transported to the sorption sites. It was found that both two- and multi-site sorption models fitted the field data equally well and the presence of surfactant enhanced the mass transfer rate of contaminant.

2.3. Multi-Component Reactive Transport Model

Finkel et al. [15] modeled surfactant and PAH migration in SEAR with a multi-component reactive transport model, which include six reactive processes: (i) surfactant micellization; (ii) surfactant sorption (formation of hemi- and admicelles); (iii) intra-particle diffusion of contaminant; (iv) contaminant sorption onto aquifer media; (v) contaminant sorption on hemi- or into admicelles; and (iv) partition of contaminant into micelles. Except the kinetic contaminant sorption, all processes were assumed to be "fast" with respect to advective transport and were described by equilibrium relationships, which were further combined with effective isotherms to determine the mass distribution between the mobile and immobile phases. To account for the slow contaminant sorption, an intra-particle diffusion approach was taken, assuming that diffusion-limited mass transfer from the bulk fluid to the intra-particle sorption sites was the dominating cause for the observed slow contaminant sorption.

An analytical model based on Fick's second law was developed, assuming that the solid grains may be approximated as sphere with sorption sites evenly distributed throughout the sphere:

$$\frac{\partial C_{jk}}{\partial t}=\frac{D_j^{app}}{r^2}\frac{\partial}{\partial r}(r^2\frac{\partial C_{jk}}{\partial r}) \tag{18}$$

where the indices j and k account for the lithological composition and grain size distribution, respectively; D_j^{app} is the apparent diffusion coefficient, accounting for the tortuosity of intra-particle pores, intra-particle porosity, and sorption in intra-particle pores. With an approximation of the transient boundary condition by a step function $c_{jk}(R_k, t) = c_{bulk}^i$ if $t_{l-1} < t < t_l$ (where R_k is the radius of grains and c_{bulk}^i is the concentration in the bulk fluid), the solute mass in a sphere of type (j,k) was given as a function of time by:

$$M_{jk}(t)=\frac{8R_k^3}{\pi}[\varepsilon_j+(1-\varepsilon_j)\cdot\rho_j K_{d,j}]\cdot\sum_{n=1}^{\infty}\frac{1}{n^2}[c_{bulk}^l - -\sum_{m=1}^{l}(c_{bulk}^{mt}-c_{bulk}^{m-1})\cdot e^{-D_j^{app}n^2\pi^2(t-t_{m-1})/R_k^2}] \tag{19}$$

where ε_j is the intra-particle porosity; ρ_j is the dry solid density; and $K_{d,j}$ is the sorption distribution coefficient in tra-particle pores. The fraction of fast and slow sorption sites were estimated from the relationship between exterior and interior surface.

In simulating column experiments, the reactive processes were included in a 1-D transport model describing advective and longitudinal dispersion, in which conservative transport and reactive processes were treated separately. The concentration profile for the reactive transport was determined with respect to the travel time, τ, or an inert tracer, and a probability density function, $g(\tau, L)$ was defined as:

$$g(\tau, L) = \frac{1}{n\sqrt{4\pi\alpha_L \tau \upsilon}} e^{-(\frac{(L-\upsilon t)^2}{4\alpha_L \tau}} \quad (20)$$

where g(τ, L) represents the distribution of arrival times of non-reactive tracer particles at the outlet of the column (length L) for a Dirac pulse input of unit mass and α_L is longitudinal dispersivity. For any fixed time, t, a reaction function, Γ(τ, t), represents a normalized concentration profile. The normalized breakthrough curve at the column outlet of a reactive tracer was represented by:

$$C(L,t) = \sum_{i=1}^{n} g(\tau_i, L)\Gamma(\tau_i, L)\Delta\tau_i \quad (21)$$

The advective-reactive transport with respect to τ was simulated to evaluate Γ(τ, t). The retarded advective transport of PAH and surfactant was governed by the corresponding effective isotherms. Transport and reactive steps were calculated sequentially coupled by mass transformation steps.

Separate contaminant and surfactant column experiments were simulated to valid the forward model, and it was found that the model could predict adequately surfactant breakthrough curves. The modeling of contaminant transport produced reasonable results, and the author suggested that the tailing of the observed breakthrough came from the diffusion-limited sorption, which was not considered in the model. However, when the column was pre-equilibrated with surfactant solutions, the model over-predicted the retardation of contaminant transport. The authors concluded that surfactant sorption and contaminant adsolubilization should be modeled as kinetic processes and added two new processes: surfactant diffusion into intra-particle macropores and contaminant diffusion to the admicelles within the intra-particle macropores. Much better fitting between simulation and experimental results was observed after such modification. Finkel et al. [15] used this model to predict the PAH and surfactant transport at a field scale, but the conclusions were not verified with field test results.

2.4. Multi-Component Multi-Phase Flow Model

Abriola et al. [1] presented a mathematical model to describe the enhanced solubilization of residual NAPLs in porous media, which incorporated the transport of surfactant, water and organic in a three-phase system — organic (o), aqueous (w), and solid (m). The mass balance equation was:

$$\frac{\partial}{\partial t}(\varepsilon_\alpha \rho^\alpha \omega_i^\alpha) + \underline{\nabla} \cdot (\rho^\alpha \omega_i^\alpha \underline{q}^\alpha) - \underline{\nabla} \cdot (\varepsilon_\alpha \rho^\alpha D_{hi}^\alpha \cdot \underline{\nabla} \omega_i^\alpha) = \sum_{\beta \neq \alpha} E_i^{\alpha\beta} \quad (22)$$

where ω_i^α is the mass fraction of component i (i = 1, …, n_c) in phase α (α = 0, w, m); ε_α is the volume fraction of the α phase; D_{hi}^α is the hydrodynamic dispersion tensor for component i in the α phase; $\underline{q}^{\alpha}$ is the Darcy velocity of the α phase; ρ^{α} is the density of the α phase; $\varepsilon_i^{\alpha\beta}$ is tge exchange of mass of component i between α and β phases; and n is the number of components. The mass exchange term in the mass balance equation incorporated both sorption and liquid-liquid inter-phase mass transfer. The non-advective flux was represented by Fick's law and species conservations were assumed in the system.

The mass balance of all species in a sphere was:

$$\frac{\partial}{\partial t}(\varepsilon_{\alpha}\rho^{\alpha}) + \underline{\nabla}\cdot(\rho^{\alpha}q^{\alpha}) = E_i^{\alpha} \tag{23}$$

where $E^{\alpha} = \sum_{i}\sum_{\beta\neq\alpha} E_i^{\alpha\beta}$

The conservative form of the component transport was:

$$\varepsilon_{\alpha}\rho^{\alpha}\frac{\partial\omega_i^{\alpha}}{\partial t} + \rho^{\alpha}\underline{q}^{\alpha}\cdot\underline{\nabla}\omega_i^{\alpha} - \underline{\nabla}\cdot(\varepsilon_{\alpha}\rho^{\alpha}D_{hi}^{\alpha}\cdot\underline{\nabla}\omega_i^{\alpha}) = \sum_{\beta\neq\alpha} E_i^{\alpha\beta} - \omega_i^{\alpha}E_i^{\alpha} \tag{24}$$

Equations (23) and (24) are subjected to constraints: $\sum_{i}\omega_i^{\alpha} = 1$ and $\sum_{\alpha}\varepsilon_{\alpha} = 1$

Darcy's law was expressed as:

$$\underline{q}^{\alpha} = -\frac{kk_{r\alpha}}{\mu_{\alpha}}\cdot(\underline{\nabla}P^{\alpha} - \rho^{\alpha}\underline{g}) \tag{25}$$

where k is the intrinsic permeability tensor; k_{ra} is the relative permeability to the α phase; P^{α} is the α phase pressure; $\underline{g}$ is the gravity vector; and μ_{α} is the α-phase dynamic viscosity.

Micellar solubilization was assumed to be the sole organic recovery mechanism, and the organic phase was assumed to be immobile. For a rigid porous medium, the organic phase and solid phase mass balance equation were expressed as:

$$n\frac{\partial}{\partial t}(s_o\rho^{o}) = E^{o} \tag{26}$$

$$(1-\mathrm{n})\frac{\partial}{\partial \mathrm{t}}(\rho^{m}) = E^{m} \tag{27}$$

where s_o is the organic phase saturation ($\varepsilon_o = ns_o$).

Aqueous phase was dropped out of the model since the surfactant solution Darcy velocity was controlled in experiment. The exchange term, E_i^{mw}, was represented as:

$$(1-n)\frac{\partial}{\partial t}(\rho^m \omega_i^m) = E_i^{mw} \tag{28}$$

By assuming that the organic phase was non-wetting and did not contact directly the solid phase, no mass exchange between organic and solid phase was allowed, and surface diffusion was also neglected. The final form of aqueous phase transport equation was:

$$ns_w \frac{\partial C_i}{\partial t} + \rho_b \frac{\partial Q_i}{\partial t} = \frac{\partial}{\partial x}\left(ns_w D_{hi} \frac{\partial C_i}{\partial x} \right) - q\frac{\partial C_i}{\partial x} + E_i^{wo} \tag{29}$$

where C_i is the mass concentration of species i(o,s); Q_i is the sorbed mass fraction of species i, and E_i^{wo} is the exchange of species i between the aqueous phase and the organic phase. For 1-D flow, D_{hi} was expressed as: $D_{hi} = \alpha_L \frac{q}{ns_w} + \tau D_{li}$, where D_{li} is the molecular diffusivity of species i, and τ is tortuosity factor here.

Surfactant sorption was modeled with a Langmuir isotherm:

$$Q_s = \frac{Q_{ms} K_L C_s}{1 + K_L C_s} \tag{30}$$

where Q_{ms} represents the maximum sorption capacity and K_L is the rate of adsorption divided by the rate of desorption. Contaminant sorption was assumed to have negligible impact on the column effluent concentration and Q_o was set to "0".

The organic mass exchange was expressed by a linear driving force model with the assumption of diffusion-limited mass transfer through a stagnant boundary layer. Flux of solute between the phases in a direction normal to the interface was expressed as:

$$E_i^{wo} = k_f^* \alpha_0 (C_{li} - C_i) \tag{31}$$

where k_f^* is the film mass transfer coefficient for species i cross the boundary layer; C_i and C_{li} are the concentrations of species i in the bulk phase and the interface, respectively; and α_0 is the interfacial contacting area. Substituting the equilibrium saturation concentration (C_{ei}) for C_i yielded the mass transfer expression:

$$E_i^{wo} = k_f \alpha_0 (C_{ei} - C_i) \tag{32}$$

Partitioning of surfactant into the organic phase was assumed to be negligible, and the mass transfer of water into the organic phase was also neglected, which resulted in $E^0 = E_0^{ow}$.

The above system of equations in terms of C_s, C_0 and S_0 were coupled through the explicit dependence of coefficients on saturation and the implicit dependence of equilibrium solubility on surfactant concentration. The model parameters were evaluated from previous laboratory batch and column experiments [23]. Third-type boundary conditions (constant total flux) for the organic and surfactant concentrations were implemented at the upstream boundary, and secondary-type conditions (zero dispersive flux) were implemented at the downstream end of the column. Good agreements between calibrated model simulations and experimental measurements were observed. Slight retardation in the breakthrough of contaminant was discovered and attributed to the sorption of surfactant. No comparison of predicted surfactant concentration breakthrough with experimental results was made. Effect of surfactant concentration on surfactant and contaminant concentration breakthrough curves was simulated, but they were not verified with experimental results.

2.5. UTCHEM Model

Brown et al. [6] presented a simulation of SEAR by adopting the modeling approach of UTCHEM, a simulator developed for enhanced oil recovery in which discrete mass balance equations were solved for each chemical component in the system. Mass balance equation for a component k was written as:

$$\frac{\partial}{\partial t}[n\overline{C}_k \rho_k] + \nabla \cdot [\sum_{i=1}^{n_p} \rho_k (C_{kl} u_l + \overline{D}_{kl})] = Q_k \tag{33}$$

where $\overline{C}_k$ is the total volume of component k in all phase per unit pore volume; C_{kl} is the volume concentration of component k in phase l; ρ_k is the density of pure component k; $\overline{D}_{kl}$ is the dispersive flux of component k in the l phase; n_p is the number of phases; and Q_k is the source term. The phase velocity, u_l, was described by Darcy's law:

$$u_l = -\frac{k_{rl} k}{u_l} \cdot (\nabla P_l - \gamma_l \nabla h) \tag{34}$$

where k_{rl} is the relative permeability; k is the intrinsic permeability tensor; u_l is the viscosity; γ_l is the specific weight ($\rho_l g$) of phase l; and h is the vertical coordinate.

The dispersive flux was assumed to have a Fickian form:

$$\overline{D}_{kl} = -nS_l K_{kl} \cdot \nabla C_{kl} \tag{35}$$

where S_l is the saturation of phase l. The dispersion tensor, K_{kl}, was given as:

$$K_{klij} = \frac{D_{kl}}{\tau} \delta_{ij} + \frac{\alpha_T}{nS_l} |u_l| \delta_{ij} + \frac{(\alpha_L - \alpha_T)}{nS_l} u_{li} u_{lj} |u_l| \tag{36}$$

where D_{kl} is the diffusivity of component k in phase l; δ_{ij} is the Kronecker delta; α_T is the transversal diffusivities; and u_{li} is the component of the Darcy velocity of phase l in direction i.

The continuity equation (eq. 33) was re-written in terms of aqueous phase pressure as:

$$nC_t \frac{\partial P_I}{\partial t} + \nabla \cdot k \cdot \lambda_{rTc} \nabla P_I = -\nabla \cdot \sum_{i=I}^{n_p} k \cdot \lambda_{rlc} \nabla h + \nabla \cdot \sum_{i=I}^{n_p} k \cdot \lambda_{rlc} \nabla P_{clI} + \sum_{i=I}^{n_{cv}} Q_k \quad (37)$$

where C_t is the total compressibility; λ_{rlc} is the relative mobility; λ_{rTc} is th total relative mobility; and P_{cl1} is the capillary pressure. The fluid phase was treated as incompressible, and the relative mobility was defined as $\lambda_{rlc} \equiv \frac{k_{rl}}{\mu_l}$ and $\lambda_{rTc} = \sum_{l=1}^{n_p} \lambda_{rl}$, respectively.

The imbibition and drainage relative permeability were modeled by the Corey-type function:

$$k_{r1} = k_{r1}^0 (S_{n1})^{nl} \quad (38)$$

$$k_{r2} = k_{r2}^0 (1 - S_{n1})^{n2} \quad (39)$$

$$k_{r3} = k_{r3}^0 (S_{n3})^{n3} \quad (40)$$

where the normalized aqueous phase and microemulsion saturation for drainage are

$$S_{nl} = \frac{(S_1 - S_{1r})}{(1 - S_{1r})}$$

and

$$S_{n3} = \frac{(S_3 - S_{3r})}{(1 - S_{3r})};$$

the normalized aqueous and microemulsion saturation for imbibition are

$$S_{nl} = \frac{(S_1 - S_{1r})}{(1 - S_{1r} - S_{2r})}$$

and

$$S_{n3} = \frac{(S_3 - S_{3r})}{(1 - S_{3r} - S_{2r})};$$

k_{r1}^0, k_{r2}^0 and k_{r3}^0 are relative permeability endpoints; n_1, n_2 and n_3 are the exponents for phase 1 (aqueous), phase 2 (organic) and phase 3 (microemulsion), and S_{1r}, S_{2r}, and S_{3r} are the aqueous, organic and microemulsion phase residual saturation, respectively. The influence of IFT upon relative permeability was modeled by making all the above parameters (k_{r1}^0, k_{r2}^0, k_{r3}^0, n_1, n_2, n_3, $S_{1r,}$ $S_{2r,}$, and S_{3r}) a function of capillary number.

The drainage and imbibition capillary pressure were modeled by the modified Brook-Corey function:

$$(p_b / p_c)^{\lambda d} = S_{n1} \tag{41}$$

$$(\mathrm{p_b} / P_c)^{\lambda i} = (1 - S_{n1}) \tag{42}$$

where p_b is the entry pressure and λ is a curve-fitting parameter (pore size index), which takes on values of λ_d or λ_i for drainage and imbibition, respectively.

Surfactant sorption was modeled by a Langmuir-type isotherm:

$$\hat{C}_3 = \frac{Q_{ms} K_L C_{31}}{1 + K_L C_{31}} \tag{43}$$

Q_{ms} was treated as a linear function of effective salinity in the model. Organic contaminant sorption was neglected in this model, and NAPL solubilization rate was also excluded due to limited availability of solubilization rate data. Heterogeneity was introduced by allowing the user to specify the permeability in each coordinate direction, and porosity for each grid cell, which implicitly accommodated the variability of capillary pressure.

The authors used above model to simulate a hypothetical field scale remediation against published aquifer, contaminant, and surfactant solution data. They concluded that the total remediation time can be reduced by more than one order of magnitude over traditional water flooding by SEAR, although no comparison between real word results and simulation was made.

3. Discussion

A two-step isotherm is incorporated for surfactant adsorption in the surfactant and contaminant coupled transport model, while the contaminant adsorption considers contaminant partition between micelles and aqueous phase, and between hemimicceles/admicelles and aqueous phase under non-aqueous conditions [32]. Only transport of surfactant and contaminant in aqueous solution are described in this model. As a result, it is not able to fully describe the processes in SEAR.

The solute transport coupled with two-, multi-site sorption model considers the non-equilibrium adsorption of surfactant and contaminant during their respective transport processes [33]. The adsorption of surfactant and contaminant are modeled by Langmuir and linear isotherm, respectively. This model emphasized the importance of non-equilibrium surfactant adsorption, however, the importance of surfactant adsorption may be exaggerated and it is not expected to be very useful since minimizing surfactant adsorption is one of the criteria in surfactant selection for SEAR. The effect of adsorbed surfactant on contaminant transport is modeled by abruptly changing the contaminant sorption parameters when surfactant reaches the site, which is another concern with this model. The existence of contaminant sorption onto aquifer media during the coupled surfactant and contaminant transports presented in this model is questionable. Although hydrophobic organic compound adsorption on aquifer media is well known, the hemimicelles/admicelles layer coated on the aquifer media surface will prevent this from happening, instead, contaminant adsolubilization is expected to be the major mechanism causing contaminant retardation during its transport. This model is essentially a single solute reactive transport model with time-dependent manipulated parameters. Although not representative of SEAR processes, the improvement of this model compared to the advection and dispersion model is that non-equilibrium adsorption of contaminant and surfactant are incorporated. The multi-component reactive transport model considers six reactive processes during the coupled surfactant and contaminant transport, in which the contaminant sorption, surfactant adsorption, and contaminant adsolubilization are considered to be at non-equilibrium conditions. The model's treatment of contaminant adsorption on aquifer media is in question, as has been discussed above. This model seems to be a reasonable one for micellar-solubilized contaminant transport in the subsurface, but the actual contaminant micellar solubilization processes in the NAPL source zone (i.e., the contaminant source clean up processes) are not incorporated.

The multi-component multi-phase flow model pays great attention to NAPL solubilization by surfactant solution, which is described with a linear driving force model [1]. However, the approach used in evaluating the "effective mass transfer coefficient" is in question, and is not expected to be generally applicable in SEAR (to be discussed later). Surfactant adsorption is assumed to be at equilibrium and represented by a Langmuir isotherm, while the sorption of contaminant is neglected, and no admicelles/hemimicelles formation and contaminant adsolubilization is considered. However, once surfactant adsorption is considered, the incorporation of contaminant retardation by adsolubilization seems to be a reasonable follow-up step. Slight retardation in the breakthrough of contaminant was observed and attributed to the sorption of surfactants, which might also be explained by the contaminant partitioning into hemimicells/admicelles formed by adsorbed surfactant. The capacity of this model in predicting surfactant transport, which is critical for NAPL dissolution and transport, was not verified with experimental results. Organic phase is treated as entrapped and immobile in this model, but mobilization of NAPL phase is very common in SEAR.

The UTCHEM model is the only one that considers NAPL mobilization in the subsurface, which is accomplished by tracking the migration of the middle phase microemulsion phase. Since electrolyte addition to ionic surfactant solution is a common practice, surfactant adsorption is also modeled as a function of salinity and surfactant concentration aside from assumption of equilibrium adsorption. Adsorption of contaminant is neglected, and NAPL solubilization is not included. The negligence of non-equilibrium contaminant solubilization in the surfactant solutions can underestimate contaminant removal during the remediation processes, while the negligence of surfactant non-equilibrium adsorption and contaminant adsolubilization may underestimate the initial breakthrough time of surfactant and contaminant.

Many processes can be involved in SEAR and widely different approaches have been applied in modeling them. Although almost all models performed reasonably in describing the experimental results, there are still many fundamental questions left unanswered, and better understanding of them will improve the reliability of modeling as a quantitative tool for prediction and design of SEAR.

The most important question is what the rate-limiting steps for contaminant removal are in SEAR. Is the NAPL removal rate in SEAR limited by solubilization? Under microscopic view, it is very likely that the transport of micellar-solubilized contaminant controls the local contaminant removal rate at high NAPL saturation, while further contaminant removal in the source zone is dissolution rate limited. Contaminant concentration at the maximum solubility of the surfactant solution has never been observed in the combined effluent samples collected from laboratory column and field investigations. On the other hand, even if the maximum contaminant solubility is reached locally, the limitations of laboratory and field sampling techniques may prevent the actual detection of contaminant concentration at these locations. It is likely that dissolution of NAPL by surfactant solutions is not limiting the NAPL removal in the beginning of SEAR, but it becomes more important as the NAPL residual saturation decreases. The NAPL ganglia distributed in the small soil pores are more difficult to be reached by surfactant solution with reduction in NAPL saturation, and the dissolution of such ganglia is limited by the mass transfer kinetics. Furthermore, a small fraction of contaminant can be adsorbed on or partitioned into the aquifer media and/or surfactant admicelles/hemimicelles, and removal of this fraction in the later stage of remediation is controlled by desorption.

Another question is surfactant adsorption and its impact on contaminant removal. Treatment of the sorption isotherms and the kinetics of surfactant adsorption/desorption plays an important role in the overall prediction accuracy of surfactant transport, and contaminant dissolution/transport as well. As has been discussed previously, the admicelles/hemimicelles layer on the soil matrix provides a good location for NAPL to adsorb/partition. Because of the surfactant monomer-micelle distribution equilibrium, desorption of admicelles/ hemimicelles and consequently the adsolubilized contaminant from the soil matrix is unlikely unless the surfactant concentration in the groundwater drops below its CMC. High surfactant concentrations are usually applied in SEAR (on the order of 1000 times of CMC), and it is expected that desorption of admicelles/hemimicelles from soil matrix will not occur during SEAR, which means that the removal of adsolubilized contaminant will be rather slow.

The importance of NAPL dissolution rate in SEAR and its representation in modeling is another major question. NAPL dissolution rate is important if dissolution is controlling the contaminant removal. Most models do not consider NAPL dissolution rate due to lack of data. The film transfer rate, interfacial area, contaminant solubilization capability of solution, and contaminant concentration in surfactant solution all affect the NAPL dissolution rate. Although contaminant solubilization capacity and contaminant concentration can be easily accounted for in SEAR, no direct relationship between interfacial area and NAPL saturation has been proposed. When the NAPL residual saturation decreases, although the total mass of NAPL decreases, the left NAPL ganglia has much greater surface area/volume ratios because of their smaller sizes. At the same time, the reduction in NAPL saturation increases the local permeability to surfactant solution, resulting in the preferential flows of surfactant solutions through the “less” contaminated regions. Although the surfactant solution is still dissolving

NAPL in the higher NAPL saturation regions, the locally low flow rate limits the efficient utilization of the solution's solubilization capacity.

Finally, the impact of interfacial tension on contaminant solubilization rate should be considered in SEAR modeling. Lower IFT means much less energy is required to create additional interface, and decreased IFT has been correlated with increased contaminant solubilization capacity and interfacial contacting area. Thus, the overall contaminant solubilization rate into the surfactant solution is expected to increase with decreasing IFT. During the transport of surfactant solution, the interracial tension properties of the surfactant solution also vary with changes in local surfactant and electrolyte concentrations, which will affect the corresponding contaminant solubilization rate. The effect of IFT on the solubilization rate has not been considered in the models developed so far, although UTCHEM considers the effect of IFT on changes in capillary pressure and permeability [6].

4. Conclusions and Recommendations

Processes including surfactant micellization, surfactant and contaminant sorption, contaminant micellar-solubilization, contaminant adsolubilization, and advective and dispersive transport of surfactant and contaminant, reduction of contaminant residual saturation, and alteration of capillary pressure may be involved in SEAR with different contributions to overall contaminant removal. SEAR modeling is further complicated by the kinetics of reactive processes. Surfactant and contaminated coupled transport model, solute transport coupled with two-, multi-site sorption model, multi-component reactive transport model, multi-component multi-phase flow model, and UTCHEM model have been developed to simulate surfactant and contaminant transport in SEAR. Nonetheless, none of these models adequately addresses all the fundamental processes due to various limitations.

Different model treatments have been adopted and agreement with experimental results is observed in each case. The actual processes that limit the concentration removal during SEAR, especially during the later stage, are still not well understood. Investigation of the fundamental processes involved in SEAR and quantification of their impacts on contaminant removal are necessary to improve the understanding of SEAR and to provide insight for design of SEAR. It is recommended that the individual process involved in SEAR be characterized separately with controlled experiences to examine its contribution to the overall contaminant removal. Identification of the processes involved in SEAR and quantification of their impacts will also improve the reliability of modeling as tool to design SEAR and make predictions.

This manuscript is reviewed by:
Kevin Chen, Ph.D.
Assistant Professor
Department of Chemical/Biomedical Engineering
FAMU-FSU College of Engineering
2525 Pottsdamer Street, Rm B337
Tallahassee, FL 32310-6046

REFERENCES

[1] Abriola, L. M., Dekker, T. J. and Pennell, K. D. (1993). Surfactant-enhanced solubilization of residual dodecane in soil columns.2. Mathematical modeling. *Environmental Science and Technology*. 27:2341-2351.

[2] Al-Futaisi, A. and Patzek, T. W. (2004). Secondary imbibition in NAPL-invaded mixed-wet sediments. *Journal of Contaminant Hydrology*. 74:61-81.

[3] Ayirala, S. C., Vijapurapu, C. S. and Rao, D. N. (2006). Beneficial effects of wettability altering surfactants in oil-wet fractured reservoirs. *Journal of Petroleum Science and Engineering*. 52:261-274.

[4] Boyd, G. R., Li, M. H., Husserl, J. and Ocampo-Gomez, A. M. (2006). Dip-angle influence on areal DNAPL recovery by co-solvent flooding with and without pre-flooding. *Journal of Contaminant Hydrology*. 82:319-337.

[5] Broholm, K., Feenstra, S. and Cherry, J. A. (2005). Solvent release into a sandy aquifer. 2. Estimation of DNAPL mass based on a multiple-component dissolution model. *Environmental Science and Technology*. 39:317-324.

[6] Brown, C. L., Pope, G. A., Abriola, L. M. and Sepehrnoori, K. (1994). Simulation of surfactant-enhanced aquifer remediation. *Water Resources Research*. 30:2959-2977.

[7] Butler, G. W., Jackson, R. E., Pickens, J. F. and Pope, G. A. (1995). An interwell solubilization test for characterization of nonaqueous-phase liquid zones. *Surfactant-Enhanced Subsurface Remediation*. 594:201-215.

[8] Cheng, H. and Reinhard, M. (2006). Measuring hydrophobic micropore volumes in geosorbents from trichloroethylene desorption data. *Environmental Science and Technology*. 40:3595-3602.

[9] Cheng, H. and Reinhard, M. (2008). The rate of 2,2-dichloropropane transformation in mineral micropores: Implications of sorptive preservation for fate and transport of organic contaminants in the subsurface. *Environmental Science and Technology*. in press:DOI: 10.1021/ es072888h (2008).

[10] Cheng, H. and Reinhard, M. (2006). Sorption of trichloroethylene in hydrophobic micropores of dealuminated Y-zeolites and natural minerals. *Environmental Science and Technology*. 40:7694-7701.

[11] Cheng, H. and Sabatini, D. A. (2007). Separation of organic compounds from surfactant solutions: A review. *Separation Science and Technology*. 42:453-475.

[12] Cheng, H., Sabatini, D. A. and Kibbey, T. C. G. (2001). Solvent extraction for separating micellar-solubilized contaminants and anionic surfactants. *Environmental Science and Technology*. 35:2995-3001.

[13] Childs, J., Acosta, E., Annable, M. D., Brooks, M. C., Enfield, C. G., Harwell, J. H., Hasegawa, M., Knox, R. C., Rao, P. S. C., Sabatini, D. A., Shiau, B., Szekeres, E. and Wood, A. L. (2006). Field demonstration of surfactant-enhanced solubilization of DNAPL at Dover Air Force Base, Delaware. *Journal of Contaminant Hydrology*. 82:1-22.

[14] Farrell, J., Grassian, D. and Jones, M. (1999). Investigation of mechanisms contributing to slow desorption of hydrophobic organic compounds from mineral solids. *Environmental Science and Technology*. 33:1237-1243.

[15] Finkel, M., Liedl, R. and Teutsch, G. (1999). Modeling surfactant-enhanced remediation of polycyclic aromatic hydrocarbons. *Environmental Modeling and Software.* 14:203-211.

[16] Fountain, J. C., Waddellsheets, C., Lagowski, A., Taylor, C., Frazier, D. and Byrne, M. (1995). Enhanced removal of dense nonaqueous-phase liquids using surfactants: Capabilities and limitations from field trials. *Surfactant-Enhanced Subsurface Remediation.* 594:177-190.

[17] John, W. W., Bao, G. B., Johnson, W. P. and Stauffer, T. B. (2000). Sorption of nonionic surfactant oligomers to sediment and PCE DNAPL: Effects on PCE distribution between water and sediment. *Environmental Science and Technology.* 34:672-679.

[18] Kent, B. and Mosquera, G. C. B. (2001). Remediation of NAPL-contaminated aquifers: Is the cure worth the cost? *Journal of Environmental Science and Health Part A: Toxic/Hazardous Substances and Environmental Engineering.* 36:1559-1569.

[19] Kim, J. and Corapcioglu, M. Y. (2003). Modeling dissolution and volatilization of LNAPL sources migrating on the groundwater table. *Journal of Contaminant Hydrology.* 65:137-158.

[20] Kim, J. and Corapcioglu, M. Y. (2001). Sharp interface modeling of LNAPL spreading and migration on the water table. *Environmental Engineering Science.* 18:359-367.

[21] Kram, M. L., Keller, A. A., Rossabi, J. and Everett, L. G. (2001). DNAPL characterization methods and approaches, part 1: Performance comparisons. *Ground Water Monitoring and Remediation.* 21:109-123.

[22] Park, S. K. and Bielefeldt, A. R. (2003). Equilibrium partitioning of a non-ionic surfactant and pentachlorophenol between water and a non-aqueous phase liquid. *Water Research.* 37:3412-3420.

[23] Pennell, K. D., Abriola, L. M. and Weber, W. J. (1993). Surfactant-enhanced solubilization of residual dodecane in soil columns.1. Experimental investigation. *Environmental Science and Technology.* 27:2332-2340.

[24] Pignatello, J. J. and Xing, B. S. (1996). Mechanisms of slow sorption of organic chemicals to natural particles. *Environmental Science and Technology.* 30:1-11.

[25] Pokrajac, D. and Deletic, A. (2006). Experimental study of LNAPL migration in the vicinity of a steep groundwater table. *Soils and Foundations.* 46:271-280.

[26] Qin, X. S., Huang, G. H., Chakma, A., Chen, B. and Zeng, G. M. (2007). Simulation-based process optimization for surfactant-enhanced aquifer remediation at heterogeneous DNAPL-contaminated sites. *Science of the Total Environment.* 381:17-37.

[27] Sahoo, D., Smith, J. A., Imbrigiotta, T. E. and Mclellan, H. M. (1998). Surfactant-enhanced remediation of a trichloroethene-contaminated aquifer. 2. Transport of TCE. *Environmental Science and Technology.* 32:1686-1693.

[28] Sauer, T. C. and Costa, H. J. (2003). Fingerprinting of gasoline and coal tar NAPL volatile hydrocarbons dissolved in groundwater. *Environmental Forensics.* 4:319-329.

[29] Schaerlaekens, J., Mertens, J., Van Linden, J., Vermeiren, G., Carmeliet, J. and Feyen, J. (2006). A multi-objective optimization framework for surfactant-enhanced remediation of DNAPL contaminations. *Journal of Contaminant Hydrology.* 86:176-194.

[30] Schluep, M., Galli, R., Imboden, D. M. and Zeyer, J. (2002). Dynamic equilibrium dissolution of complex nonaqueous phase liquid mixtures into the aqueous phase. *Environmental Toxicology and Chemistry.* 21:1350-1358.

[31] Schubert, M., Pena, P., Balcazar, M., Meissner, R., Lopez, A. and Flores, J. H. (2005). Determination of radon distribution patterns in the upper soil as a tool for the localization of subsurface NAPL contamination. Radiation Measurements 40:633-637.

[32] Setarge, B., Danzer, J., Klein, R. and Grathwohl, P. (1999). Partitioning and interfacial tracers to characterize non-aqueous phase liquids (NAPLs) in natural aquifer material. *Physics and Chemistry of the Earth Part B: Hydrology Oceans and Atmosphere.* 24:501-510.

[33] Smith, J. A., Sahoo, D., McLellan, H. M. and Imbrigiotta, T. E. (1997). Surfactant-enhanced remediation of a trichloroethene-contaminated aquifer.1. *Transport of triton X-100. Environmental Science and Technology.* 31:3565-3572.

[34] Taylor, P. 1996. The effect of an anionic surfactant on the rheology and stability of high volume fraction O/W emulsion stabilized by PVA. *Colloid and Polymer Science.* 274:1061-1071.

[35] White, M. D. and Oostrom, M. (1998). Modeling surfactant-enhanced nonaqueous-phase liquid remediation of porous media. *Soil Science.* 163:931-940.

[36] Yan, L., Thompson, K. E., Valsaraj, K. T. and Reible, D. D. (2003). In-situ control of DNAPL density using polyaphrons. *Environmental Science and Technology.* 37:4487-4493.

[37] Zhang, Y. Y., Shariati, M. and Yortsos, Y. C. (2000). The spreading of immiscible fluids in porous media under the influence of gravity. *Transport in Porous Media.* 38:117-140.

[38] Zhao, Z. K., Li, Z. S., Qiao, W. H. and Cheng, L. B. (2006). Dynamic interfacial tension between crude oil and octylmethylnaphthalene sulfonate solution. *Energy Sources Part a-Recovery Utilization and Environmental Effects.* 28:1397-1403.

[39] Zoller, U. and Rubin, H. (2001). Feasibility of in situ NAPL-contaminated aquifer bioremediation by biodegradable nutrient-surfactant mix. *Journal of Environmental Science and Health Part a-Toxic/Hazardous Substances and Environmental Engineering.* 36:1451-1471.

In: Porous Media: Heat and Mass Transfer… ISBN: 978-1-60692-437-2
Editors: J. L. Acosta and A.F. Camacho

Chapter 6

LOADING FACTOR DETERMINATION FOR GASES TRANSPORTED ACROSS MESO-POROUS ADSORBENT MEMBRANES

M.R. Othman [*] ***and Martunus***

School of Chemical Engineering, Universiti Sains Malaysia,
14300 Nibong Tebal, Penang, Malaysia

ABSTRACT

Porous inorganic membranes offer profound potential in industrial applications for gas separation in a high temperature catalytic reaction. They are also potentially useful for the separation of trace contaminants such as ammonia, hydrogen sulfide and carbonyl sulfide from coal-gasifier fuel gases. A particularly important application is in fuel cell, in which, H_2 and O_2 are used as feed gas and the membrane acts as a solid electrolyte (and/or electrode) in order to generate electricity. Other promising applications with which porous inorganic membranes have received considerable attention in a wide array of industrial operations include gas permeation for separation of CO_2 or H_2 from methane and other hydrocarbons, adjustment of H_2/CO ratio in syngas, separation of air into nitrogen and oxygen, recovery of helium and methane from bio-gas. Gas permeation involves feeding of gas mixture usually at a higher pressure or concentration into feed stream. Lower molecular weight gas species is separated from higher molecular weight gas species by virtue of different rate of diffusion through the membrane. Lower molecular weight species, having higher rate of diffusion diffuse faster into a much lower pressure or concentration in the permeate stream of the membrane. Permeation of gas species in micro and meso porous inorganic membrane has been known to be contributed primarily by the two main transport mechanisms. The first being, Knudsen diffusion mechanism and second is surface diffusion. The discussion on the internal transport mechanism by the former is relatively ample in the past and present literatures, but the transport mechanism due to the latter, however, appear to be meager possibly due to the inherent difficulty and inadequacy of robust techniques to measure it. In this paper, the

[*] M.R. Othman: *Phone: +604-5996426 *Fax: +604-5941013; *E-mail: chroslee@eng.usm.my

internal gas transport as a result of surface diffusion mechanism is given the necessary attention. The correlation between the surface mechanism and loading factor is also presented.

Key words: Permeability, Knudsen diffusion, porous membrane, separation factor

1. INTRODUCTION

Inorganic membranes made from ceramic materials, metal oxide or metal such as nickel particle-filled carbon membranes [1], amorphous alloy [2], MFI alumina membranes [3], zeolite [4] and silica membrane [5] have attracted great interest in the industry and academia due to their promising prospects for useful applications in the separation of trace contaminants such as carbon dioxide, carbon monoxide, ammonia, hydrogen sulfide and carbonyl sulfide from coal-gasifier fuel gases [6]. A particularly important application is in fuel cell, in which, H_2 and O_2 are used as feed gas and the membrane acts as a solid electrolyte (and/or electrode) in order to generate electricity [7]. Other promising applications with which porous inorganic membranes have received considerable attention in a wide array of industrial operations include gas permeation for separation of CO_2 or H_2 from methane and other hydrocarbons, adjustment of H_2/CO ratio in syngas, separation of air into nitrogen and oxygen, recovery of helium and methane from bio-gas [6].

In the application of membrane for gas separation or enrichment, the gas to be treated generally, is allowed to flow through the membrane from a more concentrated, higher pressure stream (higher driving forces) into a less concentrated, lower pressure (lower driving forces) permeate region. Separation of gas is accomplished by the difference in the molecular weight of the gas. Lower molecular weight gas species is separated from higher molecular weight gas species by virtue of different rate of diffusion in the porous media. Lower molecular weight species, having higher rate of diffusion will diffuse faster from the feeding stream into a much lower pressure or concentration in the permeate stream [8], allowing it to be captured and collected immediately.

In porous membrane for gas separation, the term *permeability* of a membrane refers to the capability of the membrane pores to transmit gas. It is a measure of the degree to which the gas molecules can flow through the membrane readily. In general, the permeability of porous inorganic membrane can be described as,

$$K = \frac{q_p \, t_m}{A_m \, \Delta P} \tag{1}$$

where, K is permeability of membrane, q_p is the gas flow rate from the permeate stream, t_m is the membrane thickness, A_m is the membrane surface area, and ΔP is the pressure difference between feed and permeate stream. This is the expression often employed to obtain laboratory measured permeability.

From the general permeability equation above, while keeping the other parameters constant, K increases when q_p increases. This relationship is true since the equation indicates that the more gas flow through the membrane pores, the higher the capability of the membrane to transmit the gas. However, for the relationship in which K increases when t_m increases as indicated in the permeability equation above, this perspective is hard to grasp. Common sense tells that the membrane is capable of transmitting the gas if the membrane is thinner. In other words, the greater the thickness of the membrane, the more difficult the gas will flow through the membrane because the gas molecules are forced to travel longer distance, which is on the contrary to the relationship given by the equation. However, in practice, the actual flow path, t, that the molecules travel is not a straight line but longer than t_m due the structural heterogeneity of the pores and meandering paths through which the gas molecules flow in order to reach the permeate stream. As we shall see later in the discussion, permeability has inverse relationship with the tortuous factor, τ. The tortuous factor is the ratio of t to t_m. Thus, increase in t_m causes the decrease in the tortuous factor, leading to the increase in the overall permeability value.

The permeability of membrane has inverse relationship with ΔP and membrane area. At higher pressure, permeability is reduced as a result of shorter mean free path of the gas molecules and also reduction of their molecular velocity when present inside the pore. In the case of membrane area, the equation suggests that larger membrane area yields lower permeability value than smaller membrane area. The relationship between permeability and membrane area can be explained as follows. Imagine a constant volume of gas flowing across small and large surface area. More gas molecules obviously occupies pores of smaller membrane area than pores of larger membrane area because the volume of gas in larger membrane area is more widely distributed than that in smaller membrane area. Since more gas molecules are present in the pores of smaller membrane area, permeability per unit area of smaller membrane area will be higher than permeability per unit area of larger membrane area.

2. Transport Mechanisms in Porous Membrane

2.1. Viscous Flow

For viscous or bulk flow mechanism in a pore, consider the flow of gas inside a cylindrical shape capillary. For deriving the viscous flow model inside a membrane pore, a cylindrical shell of thickness Δr, and length t, is selected as illustrated in figure 1 [9],

Few assumptions are made in that the mean free path of the molecules is small compared to the pore diameter of the membrane, the pore is assumed to be of a perfectly cylindrical shape, the process that occurs in the system is assumed to be in a steady state, laminar flow without the gravitational influence and that there is no molecule slip at the pore wall, implying that the gas molecules are immobile on the pore surface. Should molecules become mobile on the surface of the pore, the mobility and adsorption of gas molecules are accounted for in another transport mechanism called surface diffusion.

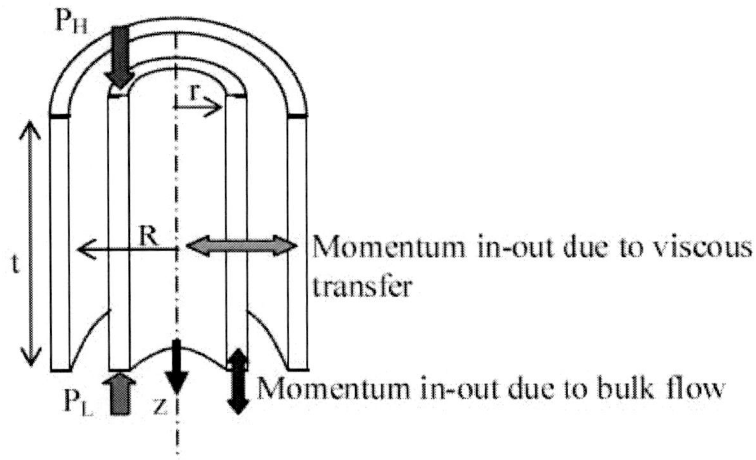

Figure 1. Various forces acting on a cylindrical pore.

Force is a product of the cross sectional area and the pressure. Since pressure has the same unit as momentum, similarly, force can be computed by multiplying area by the momentum. The contribution of various forces from and to the system is listed below:

1. Force-in, $F1$, across the cylindrical surface at r by viscous transfer

$$\begin{aligned} F1 &= (Area).(momentum\ in\ rz\ direction) \\ &= (2\pi rt).(\iota_{rz})\big|_{r} \end{aligned}$$

2. Force-out, $F2$, across the cylindrical surface at $r + \Delta r$ by viscous transfer

$$\begin{aligned} F2 &= -(Area).(momentum\ in\ rz\ direction) \\ &= -(2\pi rt).(\iota_{rz})\big|_{r+\Delta r} \end{aligned}$$

3. Force-in, $F3$, across annular surface at $z=0$ by bulk flow

$$\begin{aligned} F3 &= (Area).(\Pr essure) \\ &= (2\pi r\Delta r).(v_z^{\,2}\rho)\big|_{z=0} \end{aligned}$$

4. Force-out, $F4$, across annular surface at $z=t$ by bulk flow

$$\begin{aligned} F4 &= -(Area).(\Pr essure) \\ &= -(2\pi r\Delta r).(v_z^{\,2}\rho)\big|_{z=t} \end{aligned}$$

5. Effect of the gravity force, $F5$, on the cylindrical surface

$$F5 = (mass).(acceleration\ due\ to\ gravity)$$
$$= (2\pi r\Delta r t\rho).(g)$$

6. Applied force, *F*6, acting on annular surface at *z=0*

$$F6 = (Area).(applied\ pressure\ in\ the\ feed\ side)$$
$$= (2\pi r\Delta r).(P_H)$$

7. Applied force, *F*7, acting on annular surface at *z=t*

$$F7 = -(Area).(applied\ pressure\ in\ the\ permeate\ side)$$
$$= -(2\pi r\Delta r).(P_L)$$

For a steady flow condition, $\Sigma F = F1+F2+F3+F4+F5+F6+F7=0$

$$\left(2\pi r t\,\iota_{rz}\big|_r - 2\pi r t\,\iota_{rz}\big|_{r+\Delta r}\right) + \left(2\pi r\Delta r v_z^{\,2}\rho\big|_{z=0} - 2\pi r\Delta r v_z^{\,2}\rho\big|_{z=t}\right)$$
$$+\left(2\pi r\Delta r t\rho g\right) + \left(2\pi r\Delta r P_H - 2\pi r\Delta r P_L\right) = 0$$

simplifying the above expression,

$$2\pi r t\left(\iota_{rz}\big|_r - \iota_{rz}\big|_{r+\Delta r}\right) + 2\pi r\Delta r\rho\left(v_z^{\,2}\big|_{z=0} - v_z^{\,2}\big|_{z=t}\right)$$
$$+\left(2\pi r\Delta r t\rho g\right) + 2\pi r\Delta r\left(P_H - P_L\right) = 0$$

dividing the above equation by $(2\pi r\Delta r\, t\rho g)$,

$$\frac{r\left(\iota_{rz}\big|_r - \iota_{rz}\big|_{r+\Delta r}\right)}{\Delta r} + \frac{r\rho}{t}\left(v_z^{\,2}\big|_{z=0} - v_z^{\,2}\big|_{z=t}\right) + r\rho g + \frac{r}{t}\left(P_H - P_L\right) = 0$$

rearranging,

$$\frac{r\left(\iota_{rz}\big|_{r+\Delta r} - \iota_{rz}\big|_r\right)}{\Delta r} = \frac{r\rho}{t}\left(v_z^{\,2}\big|_{z=0} - v_z^{\,2}\big|_{z=t}\right) + r\rho g + \frac{r}{t}\left(P_H - P_L\right)$$

The flow of gas perpendicular to both end of the membrane capillary is extremely small to the extent that it can be neglected. For gas flow in a capillary, the effect of the gravity can also be neglected.

$$\frac{r\left(\iota_{rz}\big|_{r+\Delta r}-\iota_{rz}\big|_{r}\right)}{\Delta r}=\frac{r}{t}\left(P_H-P_L\right)$$

By taking the limit as Δr approaches zero,

$$\lim_{\Delta r\to 0}\left(\frac{r(\iota_{rz}\big|_{r+\Delta r}-\iota_{rz}\big|_{r})}{\Delta r}\right)=\left(\frac{P_H-P_L}{t}\right)r$$

The expression on the left becomes the derivative of $(r\iota_{rz})$,

$$\frac{d}{dr}\left(r\,\iota_{rz}\right)=\left(\frac{P_H-P_L}{t}\right)r$$

integrating both sides,

$$r\,\iota_{rz}=\left(\frac{P_H-P_L}{2t}\right)r^2+c_1$$

Applying the boundary condition where ι_{rz} is finite at $r=0$, the value of c_1 must equal zero. Thus,

$$\iota_{rz}=\left(\frac{P_H-P_L}{2t}\right)r$$

From the Newton's law of viscosity,

$$\iota_{rz}=-\mu\frac{dv_z}{dr}$$

incorporating the Newton's equation, the former equation becomes

$$\frac{dv_z}{dr}=-\left(\frac{P_H-P_L}{2\,\mu\,t}\right)r$$

integrating both sides of the equation yields,

$$v_z=-\left(\frac{P_H-P_L}{4\,\mu\,t}\right)r^2+c_2$$

applying another boundary condition where $v_z = 0$ at $r=R$

$$c_2 = \left(\frac{P_H - P_L}{4\mu t}\right)R^2$$

$$v_z = -\left(\frac{P_H - P_L}{4\mu t}\right)r^2 + \left(\frac{P_H - P_L}{4\mu t}\right)R^2$$

simplifying the above equation

$$v_z = \left(\frac{P_H - P_L}{4\mu t}\right)R^2\left(1 - \frac{r^2}{R^2}\right)$$

The last equation represents the velocity profile inside the capillary tube of the porous membrane. In order to obtain the average velocity, the above equation is multiplied and divided by the cross sectional area of the pore. Since the velocity is a function of *r*, it is integrated together over the area.

$$v_{avg} = \frac{\int_0^{2\pi}\int_0^R v_z\, r\, dr\, d\theta}{\int_0^{2\pi}\int_0^R r\, dr\, d\theta}$$

$$= \frac{\int_0^{2\pi}\int_0^R \left(\frac{P_H - P_L}{4\mu t}\right)R^2\left(1 - \frac{r^2}{R^2}\right) r\, dr\, d\theta}{\int_0^{2\pi}\int_0^R r\, dr\, d\theta}$$

$$= \left(\frac{P_H - P_L}{4\mu t}\right)R^2 \frac{\int_0^{2\pi}\left(\int_0^R \left(1 - \frac{r^2}{R^2}\right) r\, dr\right) d\theta}{\int_0^{2\pi}\left(\int_0^R r\, dr\right) d\theta}$$

using the separation of variable technique and substituting the result into the average velocity equation give,

$$v_{avg} = \frac{1}{2}\left(\frac{P_H - P_L}{4\mu t}\right)R^2$$

The permeation rate is the volumetric flow rate of the gas flowing inside the pore membrane. It can be computed by multiplying the average velocity with the cross sectional area of the pore. Nevertheless, this is the permeate rate in one cylindrical pore. For permeation rate across the membrane area, the number of pores inside the membrane must be determined. Suppose, the tubular membrane has internal radius of R_m and length L_m. Dividing the membrane area by the area of the pore and then multiplying it by the membrane porosity, ε, will give the number of pores available for flow inside the membrane. Thus, the permeate flow rate, qp is expressed as follows,

$$q_p = \frac{\varepsilon \pi R_m L_m R^2}{4\mu t}(P_H - P_L)$$

replacing the flow rate term in the observed permeability expression from equation (1),

$$K = \frac{q_p t_m}{A_m \Delta P} = \frac{\varepsilon R^2 t_m}{8\mu . t}$$

By defining, $\tau = \frac{t}{t_m}$, the permeability due to viscous flow is,

$$K_v = \frac{\varepsilon R^2}{8\tau\mu} \quad (2)$$

The permeability due to viscous flow bears a unit of $cm^3.s.g^{-1}$. The permeability is a function of porosity, pore radius, tortuous factor, and viscosity. It is interesting to note that the permeability is independent of pressure. The permeability, however, is dependent of temperature since the viscosity of gas is temperature dependent.

2.2. Gas Diffusion

The permeability equation due to voiscous flow is limited to application where the mean free path of the gas molecules is smaller than the pore diameter. In this case, the collision of inter-gas molecules is more frequent than the collision between gas molecules and the pore wall. In a high temperature, low pressure region, the mean free path of gas molecules can be much larger than the pore size of membrane and collision frequency between the molecules and pore wall increases. In this situation Knudsen diffusion may prevail with the ordinary diffusion. The combination of the ordinary and Knudsen diffusion in series having porosity, ε, is called gas diffusivity, D_g, where,

$$D_g = \varepsilon\left(\frac{1}{1/D_i + 1/D_k}\right)$$

The contribution of Knudsen and ordinary diffusion to permeability variation is known as the gas diffusion effect. The flux due to gas diffusion is in the form,

$$J_g = -D_g \frac{dC_g}{dZ}$$

Substituting the D_g,

$$J_g = -\varepsilon\left(\frac{1}{1/D_i + 1/D_k}\right)\frac{dC_g}{dZ}$$

2.3. Knudsen Diffusion

In a region where collision frequency between molecules and the pore wall increases, the gas diffusion that present is known as Knudsen diffusion. The Knudsen diffusion is expressed as,

$$D_{ki} = \frac{2r_p}{3}\sqrt{\frac{8\Re T}{\pi M_i}} \qquad (3)$$

The equation is derived from the diffusion equation obtained by employing the kinetic theory of gas and Fick's law [10]. The relevant diffusion equation is,

$$D = \frac{1}{3}\lambda \bar{c}$$

where, λ is the mean free path of the gas molecules or the distance between the molecules, a molecule and pore wall before the collision occurs. $\bar{c}$ is the molecular speed of the gas obtained by evaluating the integral of the product of the distribution of speed and the Maxwell expression for the distribution of speed. The molecular speed is then expressed in the form,

$$\bar{c} = \sqrt{\frac{8\Re T}{\pi M}}$$

For the Knudsen diffusion to take effect, the mean free path of the molecule must be larger or comparable to the pore diameter, $2r_p$, of the membrane. Considering that the potential minimum collision between a gas molecule with the pore wall occurs when $\lambda=2r_p$, the effective or available mean free path, λ_e, becomes $2(r_p - r_g)$. The effective mean free path or distance for the molecule to collide with the pore wall is obtained if its own diameter is subtracted from the pore diameter. In other words, when a gas molecule is present in a pore, the average distance for the collision is reduced and the reduction is equivalent to the difference between the pore diameter and the diameter of the molecule.

In a Knudsen regime, a gas molecule moves and stays far away from each other. This provides the basis of the assumption that each gas molecule is swept out of the pore in a given pore's cross-sectional area such as shown in figure 2.

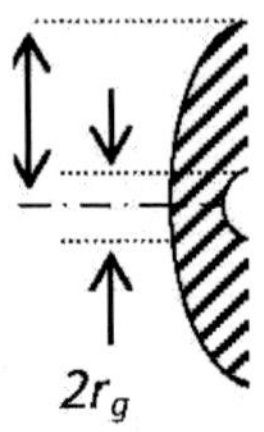

Figure 2. Collision cross-section.

If λ is substituted with the effective mean free path, λ_e, which is equal to $2(r_p - r_g)$ and rearranging the related equations, the Knudsen diffusivity in this approximation becomes,

$$D_{ki} = \frac{2(r_p - r_{gi})}{3}\sqrt{\frac{8\Re T}{\pi M_i}}; \quad for\ r_p \geq r_{gi}$$
$$D_{ki} = 0; \quad for\ r_p < r_{gi} \qquad (4)$$

The Knudsen diffusion equation above is valid for pore radius that is greater or equivalent to the molecular radius. For pore radius that is smaller than the molecular radius, the concept of Knudsen diffusion no longer holds as the mean free path of the gas molecule becomes less than the size of the pore and the fact that negative Knudsen diffusion carries no meaning. In practice, negative Knudsen diffusion can be regarded as zero or no diffusion. For a pore radius that is substantially larger than the radius of a gas molecule, equation (3) and equation (4) produce virtually similar result. However, if the pore radius is comparable to the radius of the molecule, the latter produces a result that can be significantly different from the result obtained by using the former equation.

$2r_g$

r_p

Equation (3) does not account for the contribution of pore size reduction to the variation in permeability ratio when acquiring a separation factor. This is due to the pore radius term that is cancelled out, leaving the square root of the molecular weight of the two gas species when a binary gas mixture is considered. Equation (4) in contrast, helps in explaining the increase in separation factor when membrane pore size is reduced at a molecular level since the membrane pore radius and molecular radius of the two species concerned remain intact during the calculation.

2.4. Surface Diffusion

When gas molecules are adsorbed and mobile on the pore wall, they diffuse along the pore wall in the direction of decreasing driving forces to provide substantial mobility and enhance the permeability of the gas. This additional transport is called *surface diffusion.* Employing the Fick's law to surface diffusion,

$$J_s = -D_s \frac{dC_s}{dZ}$$

D_s is the surface diffusivity measured in $m^2\ s^{-1}$. C_s is the gmole per cm^3 of gas concentration adsorbed on the pore surface. It is more convenient if the concentration can be associated with the density of the pore material (the adsorbent) so that f, which is a gram of adsorbent required to adsorb given gmoles of gas can be determined. The relationship between C_s and f is given such that,

$$C_s = \rho_m f$$

Inserting into the earlier equation gives,

$$J_s = -D_s \frac{d(\rho_m f)}{dZ}$$

$$= -D_s \rho_m \frac{df}{dZ}$$

From Henry's law for linear isotherm,

$$f \alpha kP$$

$$f = \frac{1}{z\Re T}(kP)$$

Substituting,

$$J_s = -D_s \rho_m \frac{d(\frac{kP}{z\Re T})}{dZ}$$

$$= -\frac{D_s \rho_m k}{z \Re T} \frac{dP}{dZ}$$

The above equation refers to the flux due to surface diffusion. ρ_m is the density of the membrane (g cm^{-3}), k is the Henry's adsorption constant (cm^3 g^{-1}), and D_s is the surface diffusivity (cm^2 s^{-1}), which may be estimated from an empirical equation,

$$D_S = 0.016 \exp\left(\frac{-0.45 \ \Delta H_{ads}}{\Re T}\right)$$

where,

$$\Delta H_{ads,i} = \Re T^2 \left(\frac{\partial \ln p_i}{\partial T}\right)_{ni}$$

ΔH_{ads} is the differential heat of adsorption. ΔH_{ads} carries the unit of g cm^2 s^{-2} gmol^{-1} if $\Re$=8.314×10^7 g cm^2 s^{-2} gmol^{-1} K^{-1} is used. On the other hand, ΔH_{ads} carries the unit of cal gmol^{-1} if $\Re$=1.987 cal gmol^{-1} K^{-1} is used. $pi = Pyi$ is the gas phase partial pressure of component i in equilibrium with the adsorbed phase having a specific loading of ni for that component at temperature T. The variables yi is the mole fraction of component i in the gas phase. An approximate but useful way of analyzing the differential heat of adsorption is to treat the process as condensation of vapor on a solid surface. ΔH_{ads} can be estimated using Trouton's rule and Watson correlation [11].

The pores in the membrane are neither straight nor uniform, rather, they may vary in cross-sectional areas; the paths are tortuous; and not all the void is available for the gas molecules to diffuse. In order to account for these complexities, an effective diffusion coefficient is considered. The effective diffusivity, D_e, can be obtained by applying the Fick's law and gas equation of state,

$$J_T = -D_e \frac{dC}{dZ}$$

$$J_T = -\frac{D_e}{z \Re T} \frac{dP}{dZ} \quad (5)$$

J_T is the total flux due to gas diffusion mechanisms, that includes ordinary, Knudsen, and surface diffusion. Combining all of the diffusion mechanisms yields,

$$J_T = J_g + J_s$$
$$= -\frac{1}{z\Re T}\left\{\varepsilon\left(\frac{1}{1/D_i + 1/D_k}\right) + \left(D_s\,\rho_m\,k\right)\right\}\frac{dP}{dZ}$$

comparing with equation (5),

$$D_e = \varepsilon\left(\frac{1}{1/D_i + 1/D_k}\right) + \left(D_s\,\rho_m\,k\right)$$

3. Permeability of Gas

The permeability of gas in porous media is obtained by combining all of the gas transport mechanisms discussed earlier. For viscous flow, the permeability is given by equation (2). For Knudsen and surface diffusion, the permeability can be obtained by relating the trans-membrane flux with Fick's law. The trans-membrane flux is defined as,

$$N = \frac{K}{t_m}(P_H - P_L \tag{6}$$

Equating equation (5) with equation (6), where $N = J_T$ yields,

$$\frac{K}{t_m}\left(P_H - P_L\right) = \frac{D_e}{zt\Re T}\left(P_H - P_L\right)$$

$$K = \frac{D_e}{zt\Re T}\left(P_H - P_L\right)\frac{t_m}{\left(P_H - P_L\right)} = \frac{D}{z\tau\Im} = \frac{D_e}{z\tau\Re T}$$

Substituting D_e

$$K = \frac{\varepsilon}{z\tau\Re T}\left[\left(\frac{1}{1/D_i + 1/D_k}\right) + \frac{1}{\varepsilon}\left(D_s\,\rho_m\,k\right)\right]$$

This permeability equation is attributed to diffusion mechanisms. It carries a basic unit of gmol.s.g^{-1} if common units are cancelled out. $\Re$ is the gas constant, which is equivalent to 8.314×10^{7} g cm^{2} s^{-2} gmol^{-1} K^{-1}. T is the temperature measured in Kelvin.

The permeability equation due to viscous flow and gas diffusion, also called as effective permeability, K_e, can be obtained by combining all the relevant expressions to yield,

$$K_e = \frac{\varepsilon R^2}{8\tau\mu} + \frac{\varepsilon}{z\tau\Re T}\left[\left(\frac{1}{1/D_i + 1/D_k}\right) + \frac{1}{\varepsilon}(D_s\,\rho_m\,k)\right] \quad (7)$$

Recall that the first term on the left of the equation carries the unit of $cm^3.s.g^{-1}$ whereas, the unit on the second term carries unit of $gmol.s.g^{-1}$. For consistency, a conversion factor utilizing the equation of state is used. Assuming that the pressure inside the membrane pore is equal to the average pressure in the feed and permeate stream and designating pore radius with a lower case letter, *r*, as used in a standard sign convention, the final permeability equation bearing a unit of $gmol\ s\ g^{-1}$ becomes,

$$K_e = \frac{\varepsilon}{z\tau\Re T}\left\{\frac{\overline{P}r^2}{8\mu} + \left[\left(\frac{1}{1/D_i + 1/D_k}\right) + \frac{1}{\varepsilon}(D_s\,\rho_m\,k)\right]\right\} \quad (8)$$

It can be observed that the permeability due to viscous flow increases with average pressure. For extremely small pore membrane however, the overall permeability increase is not seen because the increase due to viscous flow is offset by the diffusional effect, caused by the ordinary, Knudsen and surface diffusion. The viscous flow contribution is further reduced at higher temperature as the viscosity of the gas increases such as shown in figure 3.

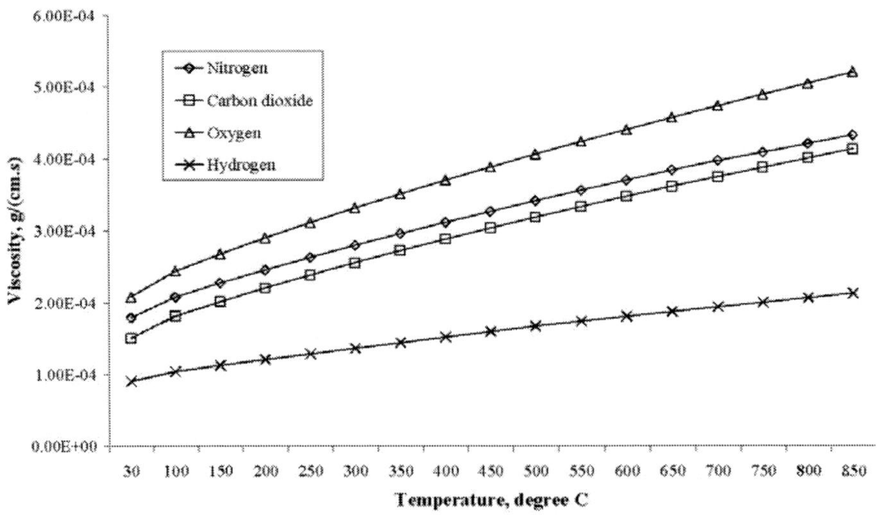

Figure 3. Viscosity of gases as a function of temperature.

Assuming that the ordinary diffusion is negligibly small compared to the Knudsen diffusion, the permeability expression for pure gas becomes,

$$K = \frac{\varepsilon}{z\,\tau\,\Re T}\left\{\frac{\overline{P}\,r^2}{8\,\mu} + \left[\frac{2(r_p - r_g)}{3}\sqrt{\frac{8\Re T}{\pi M}} + \frac{1}{\varepsilon}(D_s\,\rho_m\,k)\right]\right\}, \quad \text{for } r_p \geq r_g \tag{9}$$

For gas permeability in a binay component mixture, the mean viscosity and ordinary diffusivity are used. Knudsen diffusion in a multi-component mixture is similar to pure gas because the Knudsen diffusion of a particular gas, say gas A, is completely independent of other gases (B, C, .., *i*) in the system. When a collision does occur, the molecule of gas A collides with the pore wall, not with another molecule of its own or other gases. The permeability expression for mixed gases is,

$$K_{i\,mix} = \frac{\varepsilon}{z\,\tau\,\Re T}\left\{\frac{\overline{P}\,r^2}{8\,\mu_{mix}} + \left[\left(\frac{1}{\frac{1}{D_{i\,mix}} + \frac{1}{D_{k\,i}}}\right) + \frac{1}{\varepsilon}(D_s\,\rho_m\,k)\right]\right\}, \quad \text{for } r_p \geq r_g \tag{10}$$

For a binary mixture of gases, the viscosity value is the mean viscosity of the mixture consisting gas species A and gas species B. The mean viscosity of the mixture can be computed employing the semi-empirical formula,

$$\mu_{mix} = \sum_{i=1}^{n} \frac{x_i\,\mu_i}{\sum_{j=1}^{n} x_j\,\Phi_{i\,j}}$$

in which,

$$\Phi_{i\,j} = \frac{1}{\sqrt{8}}\left(1 + \frac{M_i}{M_j}\right)^{-\frac{1}{2}}\left[1 + \left(\frac{\mu_i}{\mu_j}\right)^{\frac{1}{2}}\left(\frac{M_j}{M_i}\right)^{\frac{1}{4}}\right]$$

Calculation of the mean viscosity for a three or more component mixture with varying composition can also be performed using the formula. Viscosity is known to be dependent on the composition of individual gas species in the mixture. In determining the ordinary diffusivity of a binary component mixture, the relevant equation that can be employed is,

$$\frac{\overline{P}\,D_{AB}}{(P_{cA}P_{cB})^{\frac{1}{3}}(T_{cA}T_{cB})^{\frac{5}{12}}\left(\frac{1}{M_A} + \frac{1}{M_B}\right)^{\frac{1}{2}}} = a\left(\frac{T}{\sqrt{T_{cA}T_{cB}}}\right)^{b}$$

For calculation of more than two component mixtures, an assumption has to be made for ease of diffusivity computation. Supposing that the diffusivity of gas species A is desired for the calculation of permeability of gas A in a multi-component gas mixture, the assumption made is that the gas species A diffuses in a pore through a stagnant non-diffusing mixture of gas species B, C, .. *i*. The ordinary diffusivity of gas species A, $D_{A\ mix}$, in the mixture becomes,

$$D_{A\,mix} = \frac{1}{\dfrac{x_B}{D_{AB}} + \dfrac{x_C}{D_{AC}} + \cdots + \dfrac{x_i}{D_{Ai}}}$$

where *i* is the individual gas species in the mixture and *x* is the mole fraction of the individual gas in the mixture. The equation can be used to determine the value of D_{AB} through D_{Ai} respectively.

For a binary gas mixture (species A and B) diffusing through a membrane, diffusivity may also be estimated from the method of Fuller. The equation to determine the diffusivity is presented in the form,

$$D_{AB} = \frac{0.00143T^{1.75}}{PM_{AB}^{1/2}\left[\left(\sum_V\right)_A^{1/3} + \left(\sum_V\right)_B^{1/3}\right]^2}$$

where,

$$M_{AB} = \frac{2}{\left(\dfrac{1}{M_A}\right) + \left(\dfrac{1}{M_B}\right)}$$

and, Σ_V is the summation of atomic and structural diffusion volumes. Several values of this property can be obtained from table 1,

The gas transport in membrane is generally, in the direction of decreasing temperature, concentration, or pressure. Controlling the boundary conditions of any of these three variables on the two sides of the membrane will provide a driving force and produce a transport of different gases through the membrane. Depending on the pore size of the membrane, gas transport by these mechanisms can vary in magnitude. The gas transport by surface diffusion, which is the focus in this chapter, is significant in a micro pore membrane (diameter of less than 2 nm), when the permeaton is conducted at low temperature and high pressure region [12]. Knudsen and surface diffusion or the combination thereof, are claimed to prevail in a meso porous membrane (pore diameter between 2 and 50 nm). Nevertheless, Knudsen diffusion was known to contribute significantly in gas transport through a membrane that exhibited the characteristic of mesoporosity [13-14].

Table 1. Diffusion Volumes

Atomic and structural diffusion volumes			
C	15.9	Fl	4.7
H	2.31	Cl	21.0
O	6.11	Br	21.9
N	4.54	I	29.8
Aromatic ring	-18.3	S	22.9
Heterocyclic ring	-18.3		
Diffusion volumes of simple molecules			
He	2.67	CO	18.0
Ne	5.98	CO_2	26.7
Ar	16.2	N_2O	35.9
Kr	24.5	NH_3	20.7
Xe	32.7	H_2O	13.1
H_2	6.12	SF_6	71.3
D_2	6.84	Cl_2	38.4
N_2	18.5	Br_2	69.0
O_2	16.3	SO_2	41.8
Air	19.7		

Since most of gas separation by membrane requires pores that exhibit microporosity structures in order to be effective in the application, the surface diffusion mechanism has become the subject of growing interest by researchers in this area of research. The effectiveness in separation is highly dependent on the affinity level of a gas for the material that makes up the membrane. In other words, the higher the adsorption of a gas on the surface of the membrane pore, the easier the gas can be transported or diffused into the permeate stream. The affinity of a gas for the membrane material is quantified by the loading factor, f, which is contained in the third term of the equation. While the loading factor can be computed from the permeabitily equation, the porosity of the membrane can be calculated using, V_p is the micro pore volume, is readily obtainable from adsorption-desorption experiment.

4. Loading Factor Determination

The membranes employed in this study in order to determine the loading factor consisted of alumina, Pd-alumina, perovskite-alumina, titania, titania-alumina, hydrotalcite and Pd. The properties of these membranes are listed in table 2. The gases that were used in the experiment were CO_2, H_2, N_2, CO, O_2, H_2/CO_2, H_2/N_2 and N_2/CO_2. The properties of the gases are given in table 3.

Table 2. Characteristics of membranes and permeability gases across the membranes

	Alumina [15]	Pd-Alumina [15]	Perovskite-Alumina[a] [16]	Titania [17]	Titania-Alumina [17]	Hydrotalcite [18-19]	Pd [20-21]	ZSM-5 [22]
Pore size, nm	3.5	7.45	3.88	2.5	1.5	4.1	4	0.5
ε	0.842	0.685	0.37	0.5	0.5	0.42	0.36	0.7
ρ_m	3.7	11.96	9	1.3	3.76	2.06	12	0.7
τ	8.44	8.37	8	3.9	5	2	0.57	4
K_{CO2}	1.543×10^{-15}	1.55×10^{-13}	B	5.1×10^{-15}	3.4×10^{-15}	1.65×10^{-14}	b	3.57×10^{-15}
K_{N2}	1.951×10^{-15}	1.18×10^{-13}	5.71×10^{-14}	6.6×10^{-15}	4.7×10^{-15}	1.485×10^{-14}	5.35×10^{-11}	2.24×10^{-15}
K_{H2}	7.352×10^{-15}	1.05×10^{-13}	B	2.14×10^{-14}	1.66×10^{-14}	b	2.02×10^{-10}	8.75 x 10-15
K_{O2}	b	B	5.46×10^{-14}	b	b	b	b	3.22×10^{-15}
K_{CO}	b	b	B	6.0×10^{-15}	4.3×10^{-15}	b	b	b
$K_{H2/N2}$	6.657×10^{-15}	1.448×10^{-13}	b	b	b	b	1.82×10^{-10}	b
$K_{H2/CO2}$	6.626×10^{-15}	1.120×10^{-13}	b	b	b	b	b	b
$K_{N2/CO2}$	b	b	b	b	b	1.555×10^{-14}	b	b
T	303	303	413	1073	1073	873	738	303
P	1.5×10^{5}	1.5×10^{5}	1.5×10^{5}	5×10^{2}	5×10^{2}	3×10^{6}	1.2×10^{5}	1.5×10^{5}

[a] $((Al_2O_3)_{0.5}(SrCo_{0.6}Fe_{0.4}O_3)_{0.5})$; b – not investigated.

Table 3. Gas moleculer properties

Gas	Molecular diameter (Å)	T_c (K)	P_c (atm)	M (g/mol)	Lennard-Jones parameters		
	[23]	[24]				σ (Å)	ε/k (K)
H_2	2.968	2.89	33.3	12.8	2.016	2.915	38
CO	3.590	3.76	133	34.5	28.01	3.590	110
N_2	3.681	3.64	126.2	33.5	28.02	3.681	91.5
O_2	-	-	154.4	49.7	32.0	3.433	113
CO_2	3.996	3.3	304.2	72.9	44.01	3.996	190

The analysis of loading factor for pure gas and gas in a binary mixture at different pressure is performed using Matlab® 2008a. The variables affecting loading factor emphasized in the analysis are the average pressure, temperature, pore size and the selected gas properties. A parameter such as gas viscosity required in the analysis can be obtained from an empirical equation utilizing Lennard-Jones parameters such as given in table 3.

The loading factor is calculated by employing the minimization of "Golden Section" search method in Matlab® 2008a. The steps of calculation and iteration are given in figure 4.

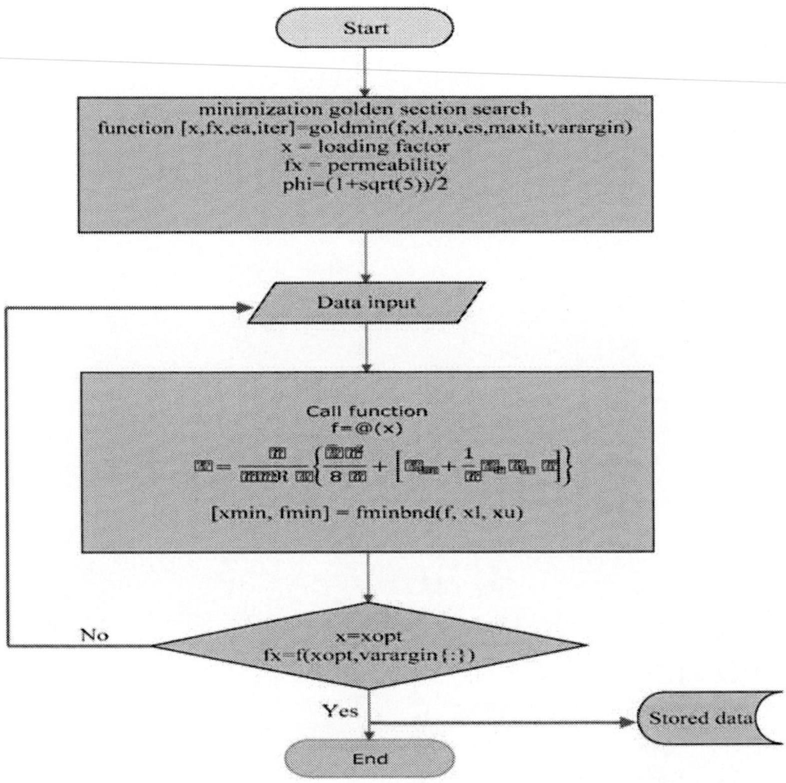

Figure 4. Algoritm for calculation of *f* with the Golden section method in Matlab.

(i) Initiating Matlab;

(ii) Applying the minimization “Golden Section” search as a function in m file (goldmin.m);

The Matlab code is given by the following,

```
goldmin.m
function [x,fx,ea,iter]=goldmin(f,xl,xu,es,maxit,varargin)
% goldmin: minimization golden section search
%[xopt,fopt,ea,iter]=goldmin(f,xl,xu,es,maxit,p1,p2,...):
% uses golden section search to find the minimum of
% input:
% f = name of function
% xl, xu = lower and upper guesses
% es = desired relative error (default =0.0001%)
% maxit = maximum allowable iterations (default =50)
% p1,p2,...= additional parameters used by f
% output:
% x = location of minimum
% fx = minimum fanction value
% ea = approximate relative error (%)
% iter = number of iterations
if nargin<3,error('at least 3 input arguments required'),end
if nargin<4|isempty(es), es=0.0001;end
if nargin<5|isempty(maxit), maxit=50;end
phi=(1+sqrt(5))/2;
iter=0;
while(1)
d = (phi-1)*(xu-xl);
x1 = xl + d;
x2 = xu -d;
if f(x1,varargin{:}) < f(x2,varargin{:})
xopt = x1;
xl = x2;
else
xopt = x2;
xu = x1;
end
iter=iter+1;
if xopt~=0, ea = (2-phi)*abs((xu-xl)/xopt)*100;end
if ea <= es | iter >= maxit,break,end
end
x=xopt;fx=f(xopt,varargin{:});
```

(iii) Collecting together the data input shown in table 1 and 2. The surface diffusion coefficient was calculated using eqs. (8) and (9);

(iv) Calling function (goldmin.m) in Matlab windows. The structure of call function for CO_2 across alumina membrane is given below for reference;

```
>> f=@(x) ((1.49E-17)+(2.02E-18)+ (x*4.18E-16));
>> [xmin, fmin] = fminbnd(f, 3.65, 20)
```

(v) Tabulating and analyzing results of steps (ii) to (iv). An example of one of the iterations is provided as follows,

```
>> options = optimset('display','iter');
>> xmin = fminbnd(f, 3.65, 20, options)
```

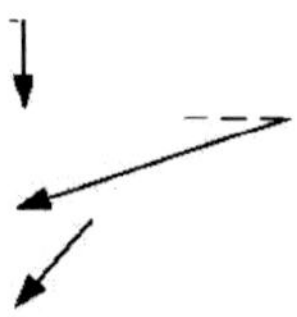

permeability eq.

x = loading factor

f = permeability

Optimization terminated:

the current x satisfies the termination criteria using OPTIONS.TolX of 1.000000e-004

xmin = 3.6501

From this study, the loading factor was the highest for Pd membrane, followed by hydrotalcite, titania-alumina, perovskite-alumina, titania, alumina and ZSM-5. The affinitie behavior of H_2 for Pd and CO_2 for hydrotalcite membrane is apparent from the high value of the respective loading factor as shown in table 4. It is also interesting to note that while N_2 is an inert gas, it appeared to exhibit a high level of affinity for hydrotalcite, Pd, Pd-alumina, perovskite-alumina and titania-alumina membranes. N_2 gas exhibited an affinity for titania, alumina and ZSM-5 membranes to a much lower extent.

Func-count	x	f(x)	Procedure
1	9.89514	4.15309e-015	Initial
2	13.7549	5.76645e-015	golden
3	7.50971	3.15598e-015	golden
4	6.03543	2.53973e-015	golden
5	5.12428	2.15887e-015	golden
6	4.56115	1.92348e-015	golden
7	4.21312	1.77801e-015	golden
8	3.99803	1.68810e-015	golden
9	3.86509	1.63253e-015	golden
10	3.78294	1.59819e-015	golden
11	3.73216	1.57696e-015	golden
12	3.70078	1.56384e-015	golden
13	3.68138	1.55574e-015	golden
14	3.66940	1.55073e-015	golden
15	3.66199	1.54763e-015	golden
16	3.65741	1.54572e-015	golden
17	3.65458	1.54453e-015	golden
18	3.65283	1.54380e-015	golden
19	3.65175	1.54335e-015	golden
20	3.65108	1.54307e-015	golden
21	3.65067	1.54290e-015	golden
22	3.65041	1.54279e-015	golden
23	3.65026	1.54273e-015	golden
24	3.65016	1.54269e-015	golden
25	3.65010	1.54266e-015	golden
26	3.65006	1.54265e-015	golden

The affinitive energy of O_2 for perovskite-alumina was much higher than ZSM-5, suggesting that the former membrane was capable of adsorbing of O_2. On the other hand, the affinity of CO for titania-alumina was much higher than titania, suggesting that the former membrane was capable of adsorbing of CO. However, if N_2 was present, there would be competitive adsorption between N_2 and CO_2 since the loading factor of two gases across perovskite-alumina membrane was of almost similar in magnitude. The loading factor for binary gases appeared the highest for Pd membrane in a mixture of two gases, followed by hydrotalcite, alumina and Pd-alumina. The affinitive behavior of H_2/N_2 mixture for Pd and CO_2/N_2 for hydrotalcite membrane is apparent from the high value of the respective loading factor as shown in table 4.

Table 4. Loading factor (*f*), mol/g

Membrane	Gas and mixture of two gases								
	CO_2	N_2	CO	O_2	H_2	H_2/CO_2	H_2/N_2	CO_2/N_2	
Alumina	3.65[c]		3.07[c]	b	b	11.712[c]	10.56	10.61	b
Pd-alumina	114[c]		86[c]	b	b	75.5[c]	105	82	b
Perovskite-alumina	b		72.3[d]	b	69.1[d]	b	b	b	b
Titania	6.32[e]		7.29[e]	6.56[e]	b	23.2[e]	b	b	b
Titania-alumina	16.8[e]		22.25[e]	20.35[e]	b	78.5[e]	b	b	b
Hydrotalcite	635[g]		551[g]	b	b	b	b	b	597
Pd	b		258[h]	b	b	980[h]	b	880	b
ZSM-5	5.7[c]		1.74[c]	b	3.04[c]	7.17[c]	b	b	b

b - not investigated

[c] T = 303 °K; [d] T = 413 °K; [e] T = 1073 °K; [g] T = 873 °K; [h] T = 738 °K.

[c] $K_{CO2} = 1.543 \times 10^{-15}$, $K_{N2} = 1.951 \times 10^{-15}$, $K_{H2} = 7.352 \times 10^{-15}$; $K_{H2/CO2} = 6.626 \times 10^{-15}$; $K_{H2/N2} = 6.657 \times 10^{-15}$ mol.cm/cm^2.s.Pa for alumina membrane.

[c] $K_{CO2} = 1.55 \times 10^{-13}$, $K_{N2} = 1.18 \times 10^{-13}$, $K_{H2} = 1.05 \times 10^{-13}$; $K_{H2/CO2} = 1.433 \times 10^{-13}$; $K_{H2/N2} = 1.120 \times 10^{-13}$ mol.cm/cm^2.s.Pa for Pd-alumina membrane.

[c] $K_{CO2} = 3.57 \times 10^{-15}$, $K_{N2} = 2.24 \times 10^{-15}$, $K_{H2} = 8.75 \times 10^{-15}$, $K_{O2} = 3.22 \times 10^{-15}$ mol.cm/cm^2.s.Pa for ZSM-5 membrane.

[d] $K_{N2} = 5.71 \times 10^{-14}$, $K_{O2} = 5.46 \times 10^{-14}$ mol.cm/cm^2.s.Pa for perovskite-alumina membrane.

[e] $K_{CO2} = 5.1 \times 10^{-15}$, $K_{N2} = 6.6 \times 10^{-15}$, $K_{H2} = 2.14 \times 10^{-14}$, $K_{CO} = 6.0 \times 10^{-15}$ mol.cm/cm^2.s.Pa for titania membrane.

[e] $K_{CO2} = 3.4 \times 10^{-15}$, $K_{N2} = 4.7 \times 10^{-15}$, $K_{H2} = 1.66 \times 10^{-14}$, $K_{CO} = 4.3 \times 10^{-15}$ mol.cm/cm^2.s.Pa for titania-alumina membrane.

[g] $K_{CO2} = 1.65 \times 10^{-14}$, $K_{N2} = 1.485 \times 10^{-14}$; $K_{CO2/N2} = 1.555 \times 10^{-14}$ mol.cm/cm^2.s.Pa for hydotalcite membrane at $P = 3 \times 10^6$.

[h] $K_{H2} = 2.02 \times 10^{-10}$, $K_{N2} = 5.35 \times 10^{-11}$; $K_{H2/N2} = 1.82 \times 10^{-10}$ mol.cm/cm^2.s.Pa for Pd membrane.

Figure 5 to 13 show the gas permeability of the gases across different types of membranes, after incorporating the loading factors from table 4. The permeabilities of all the membranes are independent of pressure, in the range from 1 x 10^5 to 3 x 10^5 Pa and at different operating temperatures. This suggests that Knudsen diffusion is the dominant internal transport mechanism of the gases across the membranes.

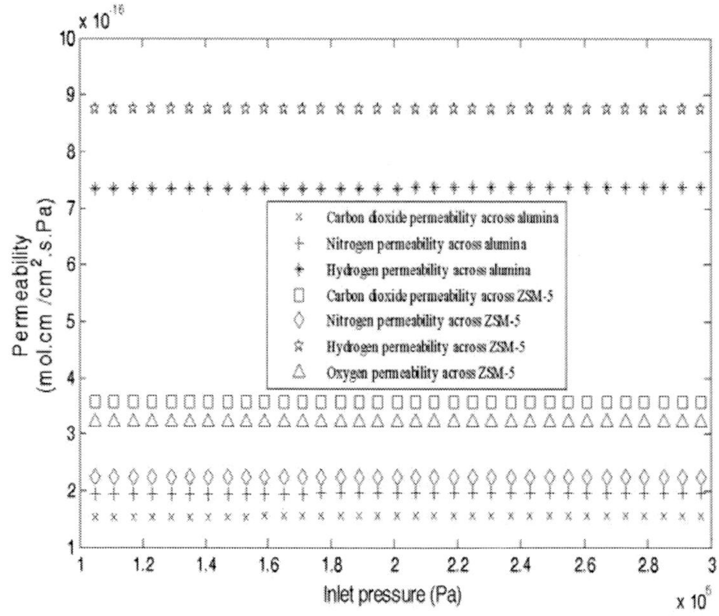

Figure 5. Permeability of H_2, N_2, CO_2 and O_2 across alumina and ZSM-5 membranes (303 K).

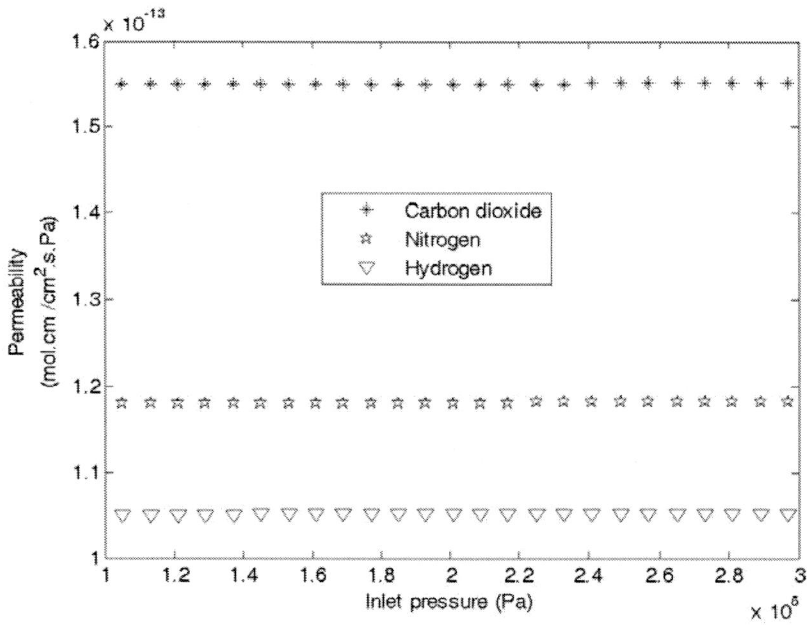

Figure 6. Permeability of H_2, N_2 and CO_2 across Pd-Alumina membrane at 303 K.

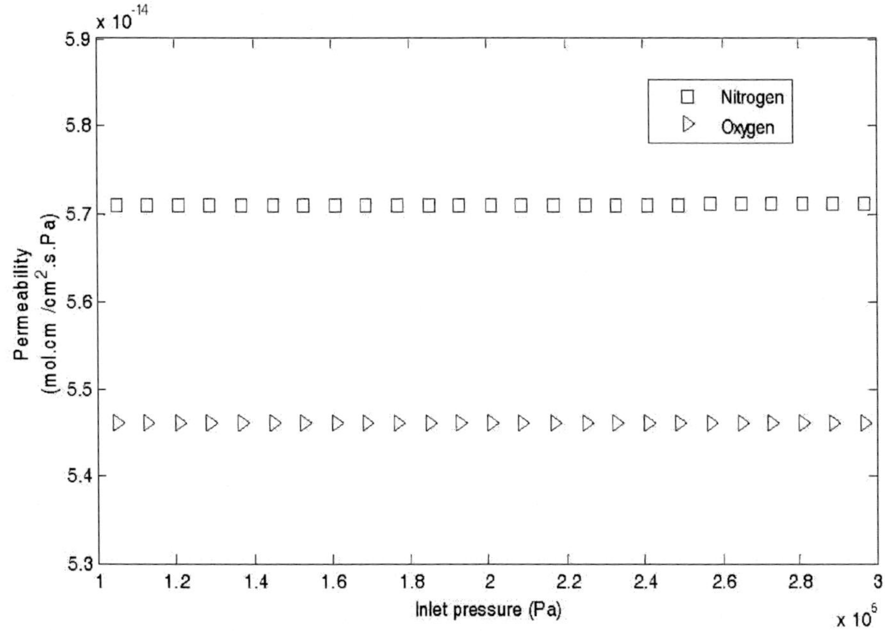

Figure 7. Permeability of N_2 and O_2 across Perovskite-Alumina membrane at 413 K.

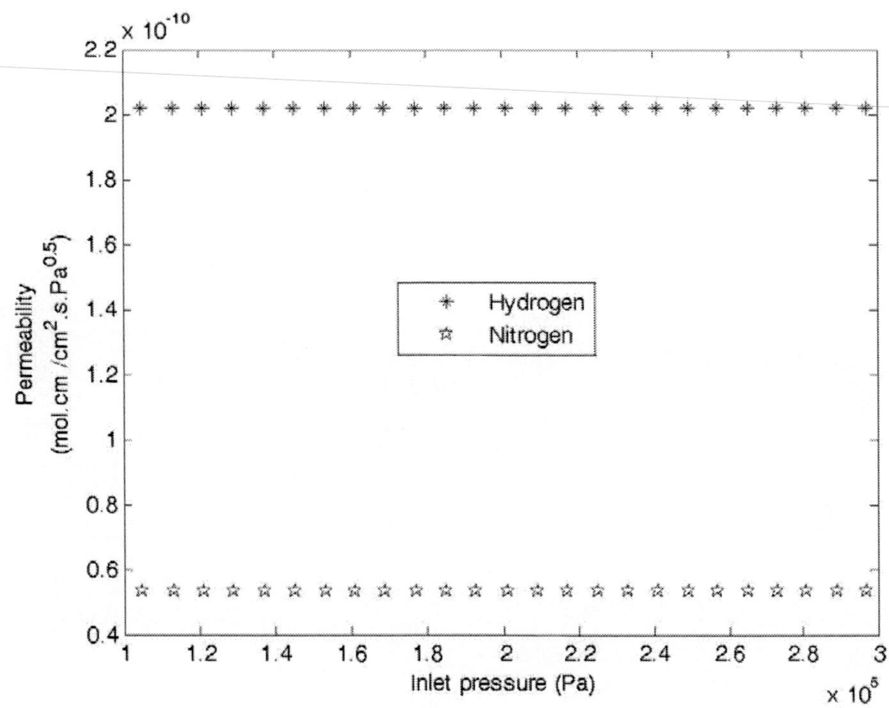

Figure 8. Permeability of H_2 and N_2 across Pd membrane at 738 K.

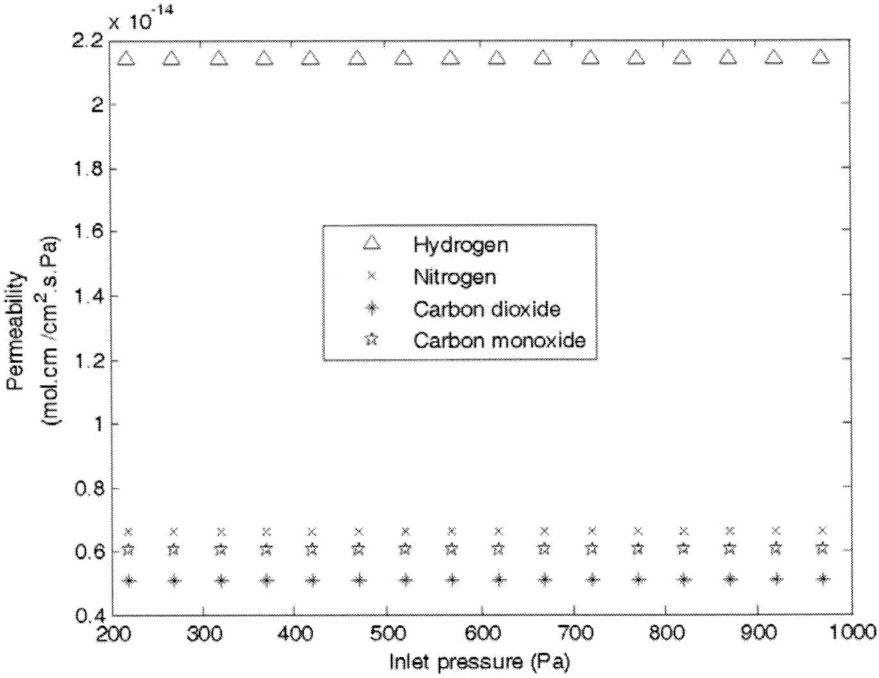

Figure 9. Permeability of H2, N_2, CO and CO_2 across Titania membrane at 1073 K.

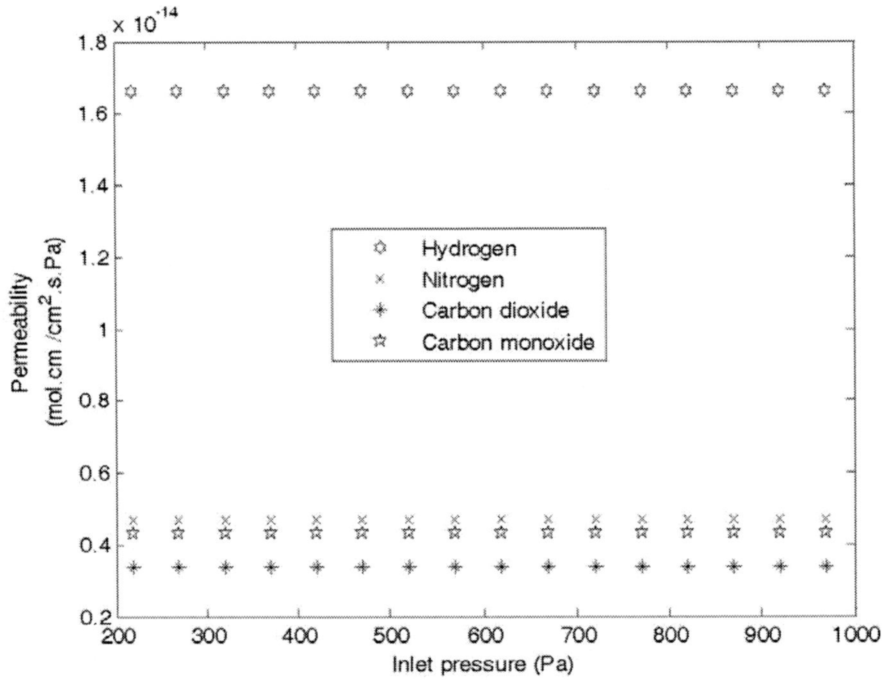

Figure 10. Permeability of H2, N_2, CO and CO_2 across titania-alumina membrane at 1073 K.

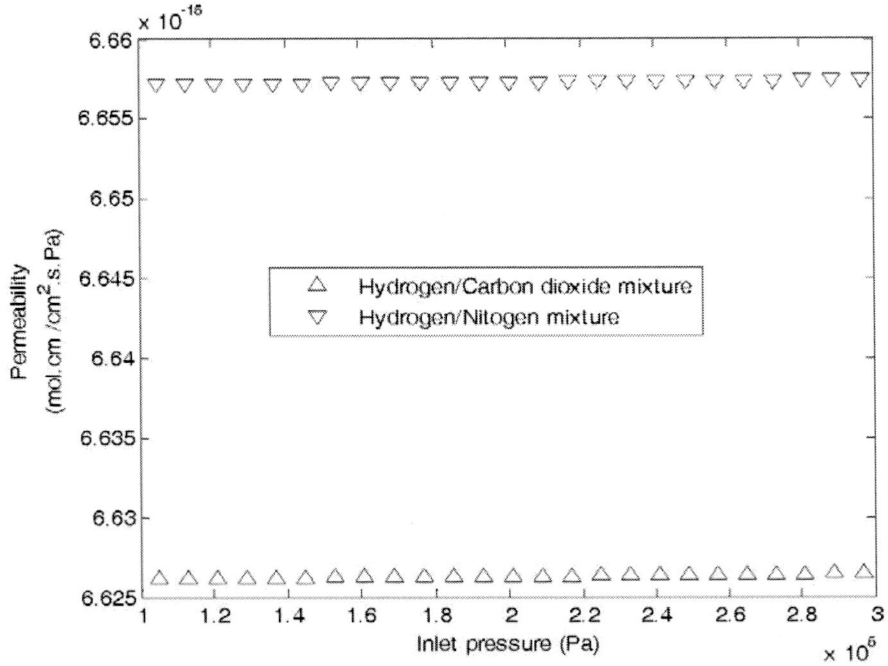

Figure 11. Permeability of H_2/CO_2 and H_2/N_2 across alumina membrane at 303 K.

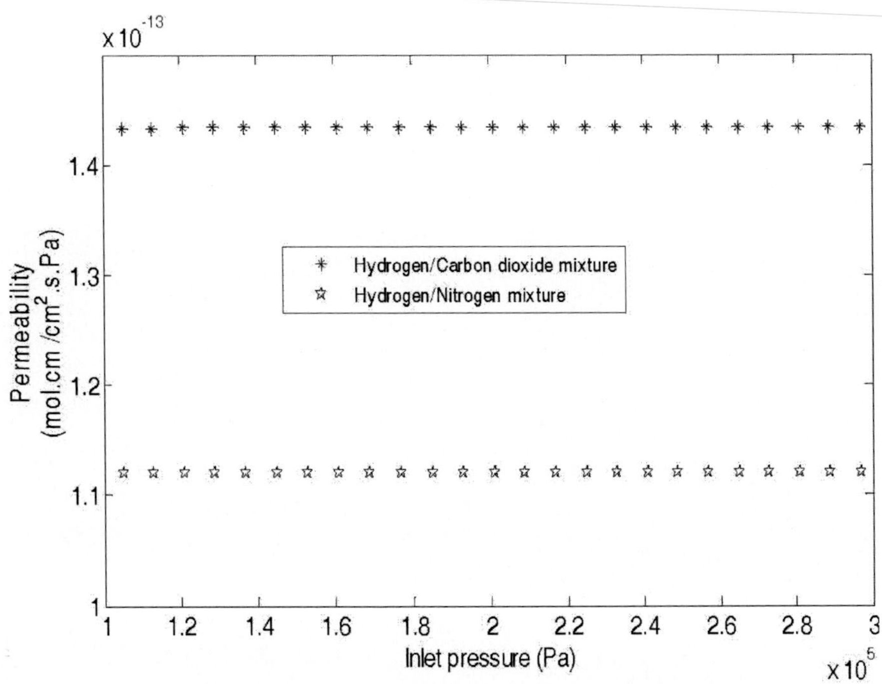

Figure 12. Permeability of H_2/CO_2 and H_2/N_2 across Pd-alumina membrane at 303 K.

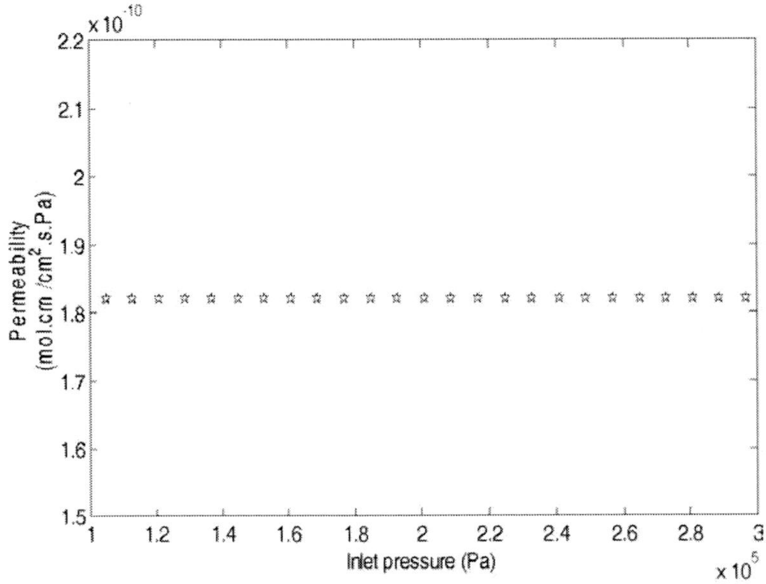

Figure 13. Permeability of H_2/N_2 across Pd-alumina membrane at 738 K.

Figures 14 and 15 show the permeabilities of gases versus inlet pressure across hydrotalcite membrane at 873 K. The gases permeabilities across hydrotalcite membrane are dependent on inlet pressure as shown by the increasing trend of permeabiliity curves, in the range 5×10^5 to 4×10^6 Pa. Since the membrane is mesoporous, the increase possibly suggests that the surface diffusion mechanism is the dominant gas transport in the membrane system. Values of loading factor and permeabilities of gases across hydrotalcite membrane at different operating pressures and 873 K are listed in table 5.

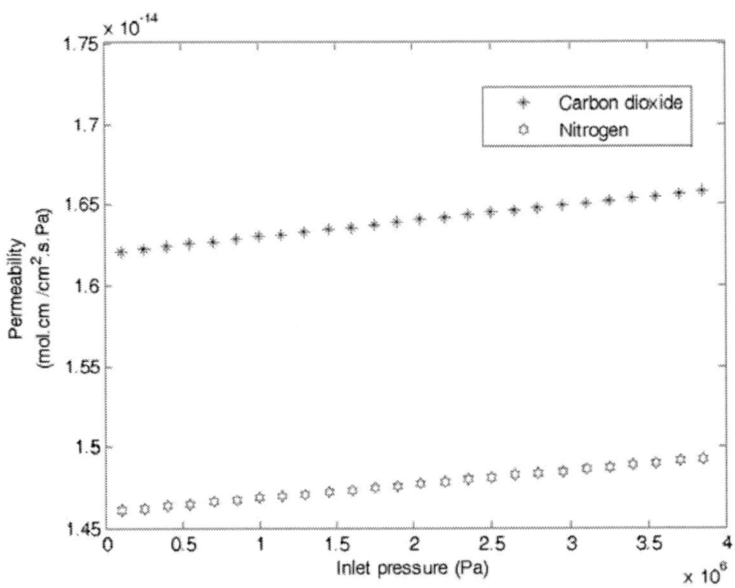

Figure 14. Permeability of N_2 and CO_2 across Hydrotalcite at 873 K.

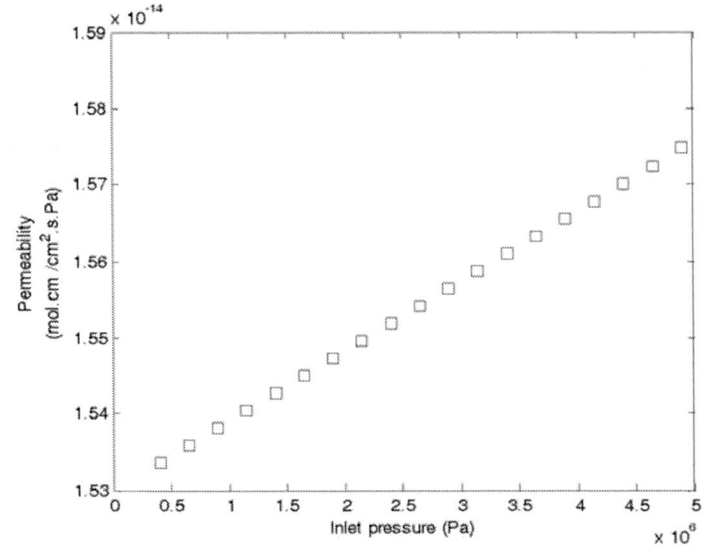

Figure 15. Permeability of CO_2/N_2 mixture across Hydrotalcite at 873 K.

Table 5. The loading factor and permeabilities at different inlet pressure*

Pressure, x 10^5 (Pa)	Permeability, mol.cm/cm^2.s.Pa			Loading factor, mol/g		
	CO_2	N_2	CO_2/N_2	CO_2	N_2	CO_2/N_2
5	1.625 x 10^{-14}	1.465 x 10^{-14}	1.534 x 10^{-14}	622.5	538.5	584.5
10	1.63 x 10^{-14}	1.469 x 10^{-14}	1.539 x 10^{-14}	625	541	587
20	1.64 x 10^{-14}	1.477 x 10^{-14}	1.548 x 10^{-14}	630	546	592
30	1.65 x 10^{-14}	1.485 x 10^{-14}	1.555 x 10^{-14}	635	551	597
40	1.66 x 10^{-14}	1.493 x 10^{-14}	1.567 x 10^{-14}	640	556	602

*Hydrotalcite membrane at T = 873 K.

NOTATIONS

A_m	Membrane surface area cm^2
D_{ij}	Ordinary diffusivity cm^2
D_k	Knudsen diffusivity cm^2.s^{-1}
D_s	Surface diffusivity cm^2.s^{-1}
f	Loading factor mol.g^{-1}
K	Permeability [mol.s^{-1}].[cm].[cm^{-2}].[g^{-1}.cm.s^2]
M	Molecular weight g.mol^{-1}

P	Average pressure in the pore $g.cm^{-1}.s^{-2}$
p	Gas phase partial pressure $g.cm^{-1}.s^{-2}$
q_p	Permeate flow rate $mol.s^{-1}$
$\Re$	Gas constant 8.314×10^{7} $g.cm^{2}.s^{-2}.mol^{-1}.K^{-1}$
r_g	Molecular radius of gas cm
r_p	Membrane pore radius cm
T	Temperature K
t_m	Membrane thickness cm
y	Mole fraction of gas -
Z	Compressibiity factor -
ΔH_{ads}	Heat of adsorption $\times10^{7}$ $g.cm^{2}.s^{-2}.mol^{-1}$
τ	Tortuous factor -
ρ_m	Density of the membrane $g.cm^{-3}$
μ	Viscosity $g.cm^{-1}.s^{-1}$
ε	Membrane porosity -

Subscript

i	Component i
j	Component j

REFERENCES

[1] L. Zhang, X. Chen, C. Zeng, Nanping Xu, *J. Membr. Sci.* 281 (2006) 429-434.

[2] Y. Shimpo, S.I. Yamaura, M. Nishida, H. Kimura, A. Inoue, *J. Membr. Sci.* 286 (2006) 170-173.

[3] S. Miachon, P. Ciavarella, L. van Dyk, I. Kumakiri, K. Fiaty, Y. Schuurman, J.A. Dalmon, *J. Membr. Sci.* 298 (2007) 71-79.

[4] X. Yin, G. Zhu, Z. Wang, N. Yue, S. Qiu, *J. Micro. Meso. Mat.* 105 (2007) 156- 162.

[5] S. Gopalakrishnan, Y. Yoshino, M. Nomura, B.N. Nair, S.I. Nakao, *J. Membr. Sci.* 297 (2007) 5-9.

[6] V.K. Venkataraman, L.K. Rath, S.A. Stern, *Key Eng. Mater.* 61:62 (1991) 347-352.

[7] R. Yegani, H. Hirozawa, M. Teramoto, H. Himei, O. Okada, T. Takigawa, N. Ohmura, N. Matsumiya, H. Matsuyama, *J. Membr. Sci.* 291 (2007) 157-164.

[8] B.H. Howard, R.P. Killmeyer, K.S. Rothenberger, A.V. Cugini, Morreale, R.M. Enick, F. Bustamante, *J. Membr. Sci.* 241 (2004) 207–218.

[9] R.B. Bird, W.E. Stewart and E.N. Lightfoot (1960). Transport Phenomena, 4th ed. New York: John Wiley & Son

[10] P.W. Atkins (1994). *Physical Chemistry*, 5th ed. p. 37, 40, 824. London: Oxford University Press

[11] R.M. Felder and R.W. Rousseau (1986). *Elem. Prin. Chem. Process*, 2nd ed. p. 364-365. New York: John Wiley & Son

[12] J. Zaman, A. Chakma. Inorganic membrane reactors. *J. Mater. Sci.* 92 (1994) 1-28.

[13] Y.K. Cho, K. Han, K.H. Lee, *J. Membr. Sci.* 104 (1995) 219-230.

[14] K.H. Lee, S.T. Hwang, *J. Colloid Interface Sci.* 110 (1985) 544-555.

[15] M.R. Othman, J. Kim, *J. Micro. Meso. Mat.* 112 (2008) 403-410.

[16] M.R. Othman, J. Kim, *J. Ind. Eng. Chem. Res.* 47 (9) (2008) 3000-3007.

[17] A.L. Ahmad, M.R. Othman, H. Mukhtar, *Int. J. Hydrogen Energy.* 29 (2004) 817-828.

[18] M.R. Othman, N.M. Rasid, W.J.N. Fernando, J. Micro. Meso. Mat. 96 (2006) 23-28.

[19] M. Sahimi, T.T. Tsotsis, Dept. of Chemical Engineering and Material Science, University of Southtern California (2006) 1-38.

[20] A. Basile, A. Criscuoli, F. Santella, H. Driolli, *Gas. Sep. Purif.* 10:4 (1996), 243-254.

[21] B.D. Morreale, M.V. Ciocco, R.M. Enick, B.I. Morsi, B.H. Howard, A.V. Cugini, K.S. Rothenberger, *J. Membr. Sci.* 222 (2003) 87-97.

[22] H.S. Oh, M.H. Kim, H.K. Rhee, *Surface Science and Catalysis.* 105 (1997) 2217-2224.

[23] D.E. Fain, G.E. Roettger, J. *Eng. Gas Turb. Power.* 115 (1993) 631-638.

[24] J.D. Way, D.L. Roberts, *Sep. Sci. Techol.* 27 (1992) 29–41.

In: Porous Media: Heat and Mass Transfer… ISBN: 978-1-60692-437-2
Editors: J. L. Acosta and A.F. Camacho

Chapter 7

GRANULAR BEDS COMPOSED OF DIFFERENT PARTICLE SIZES: EXPERIMENTAL AND CFD APPROACHES

***Ricardo P. Dias**[*] [a, b] **and Carla S. Fernandes** [c]*

[a] CEFT - Centro de Estudos de Fenómenos de Transporte,
Faculdade de Engenharia da Universidade do Porto,
Rua Dr. Roberto Frias, 4200-465 Porto, Portugal
[b] Department of Chemical and Biological Technology,
Escola Superior de Tecnologia e de Gestão,
Instituto Politécnico de Bragança, Campus de Santa Apolónia,
5301-854 Bragança, Portugal
[c] Department of Mathematics, Escola Superior de Tecnologia e de Gestão,
Instituto Politécnico de Bragança, Campus de Santa Apolónia,
5301-854 Bragança, Portugal, tel.: +351273303127,
fax: +351273313051, e-mail: cveiga@ipb.pt

ABSTRACT

The porosity, tortuosity, permeability and heat exchange characteristics from binary packings, containing mixtures of small d and large D spherical particles, are analysed in the present work. Binary packing porosity (ε), tortuosity (τ), permeability and heat exchange performance are dependent on the volume fraction of large particles, x_D, present in the mixtures, as well as on the particle size ratio, $\delta = d/D$. In the region of minimum porosity from the binary mixtures (containing spheres with diameter d and D), heat exchange performance and permeability from binary packing are higher than that of the packing containing the small particles d alone (mono-size packing). The δ region where the permeability of binary packing is higher than the permeability of mono-size packing of particles d is located in the range $0.1 \leq \delta < 1.0$. An increase in permeability by a factor of two is achieved for particle size ratios between 0.3 and 0.5. Tortuosity can be modelled by the simple function $\tau = 1/\varepsilon^{\alpha}$ and it is shown that, in the region of

[*] Ricardo P. Dias: tel.: +351273303150, fax: +351273313051, e-mail: ricardod@ipb.pt

minimum porosity, α varies between 0.5 (mono-size packing) and 0.4 (binary packing with δ close to 0.03). Due to the tortuosity increase, binary mixtures give rise to Kozeny's coefficients substantially higher than five.

Using the commercial finite element software package POLYFLOW® it was possible to confirm the heat exchange enhancement referred above. The obtained improvement on the thermal performance is related to the increase of effective thermal conductivity in the binary packing and to the increase in transversal thermal conductivity due to the porosity decrease and tortuosity increase. For non-Newtonian fluids from the power-law type, τ decreases with the decrease of the flow index behaviour

1. INTRODUCTION

During the latter half of the 19th Century widespread introduction of slow sand filtration in Europe occurred. Considerable reduction of drinking waterborne transmission of disease was achieved in communities supplied with slow sand filtered water. Rapid gravity filters were developed during the 20th Century to operate at appreciably higher filtration rates than slow sand filters, utilizing coarser media. The previous application of chemical coagulants and a clarification process was found to be needed in order to achieve drinking water of similar quality to slow sand filtered water [1].

Darcy's major contribution is undoubtedly in the area of filter hydraulics with the discovery of an empirical law that bears his name [2]. The experiments from Darcy and colleagues were performed with silica sand from Saône river, with the following composition [2, 3]: 58 percent of sand with grains smaller than 0.77 mm diameter, 13 percent of sand with 1.1 mm diameter grains, 12 percent of sand with 2 mm grains, and 17 percent of gravel and shell fragments of various sizes [2, 3].

Recently, the field of particle size distribution characterization and measurement has experienced a renaissance [4]. Packed bed hydrodynamic chromatography [5-9] had an important role in the referred revitalization, this size fractionation technique being based on the use of the parabolic profile occurring in the pores from a packed column. Large particles or molecules [5-9] migrate faster through the packed column than the smaller ones since the former particles/molecules are more excluded from the regions with low velocity close to the nonporous packing particles. Packed beds containing nonporous glass spheres of different sizes were used by Dias et al. [8] in order to analyse starch by hydrodynamic chromatography.

Slalom chromatography is giving the first steps, being used in the size fractionation of large double stranded DNA fragments. The elution order is the opposite of that from hydrodynamic chromatography, since in the former technique the larger DNA strands are eluted after the smaller ones (figure 1). In slalom chromatography, long DNA strands must turn many times around packing particles and one each turn they are exposed to a significant frictional force, this force increasing with the increase of DNA length and, therefore, larger DNA molecules being more retarded [9-13]. Due to the tortuosity increase, the use of granular beds composed by different particles sizes may increase DNA retardation [9].

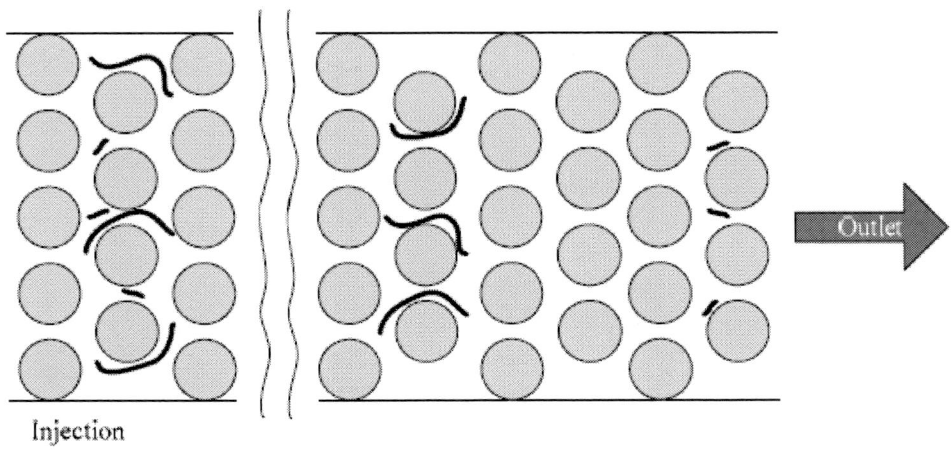

Figure 1. Slalom chromatography separation mechanism.

Trickle-bed reactors are fixed beds of catalyst contacted by cocurrent downflow of gas and liquid. They are widely used in petroleum, petrochemical, and chemical industries, pollution abatement and biochemical and electrochemical processing [13]. In order to improve the wetting of the catalyst, avoid wall effects and to provide better temperature uniformity in the reactor (particularly for highly exothermic reactions) the catalyst is usually diluted with fines (small, inert, and nonporous particles of about 0.1 catalyst diameter).

Al-Dahhan et al. (1995) [14] reported two reproducible dry packing procedures of filling the trickle-bed reactor with catalyst and fines, the final mixture presenting a good uniformity. The reactor was filled by portion mixing or whole-bed procedures. In both methods the catalyst was packed and then the fines were added to the reactor in order to fill the void volume of the skeleton formed by the catalyst particles. The ratio between the diameter of the catalyst and the diameter of the fines was large and, therefore, the fines were able to migrate to the void space by tapping or vibrating the reactor.

A good reproducibility and uniformity was also reported by Dias et al. [15] using a wet method of packing mixtures of glass spheres with two different particle sizes (binary packing). A water-glycerol was used as a binder between the different sized particles. The prepared mixture was then transferred to a square acrylic transparent column and the uniform distribution within the packing was checked by image analysis. Binary or ternary mixtures of different sized nonporous glass spheres - built by using the referred wet packing procedure - were used in studies devoted to porosity [16], tortuosity [17], permeability [18, 19] and heat exchange [19].

Granular beds greatly increase the heat transfer coefficients. However, the flow resistance in porous media is very large [20]. Dias et al. [19] shown that in the region of minimum porosity of particulate binary mixtures, heat exchange and permeability, k, were higher than the ones obtained with a mono-size packing built with the same small particles used in the binary packing. High porosity metallic foams can also be used in order to overcome the large pressure drops of granular fixed beds [21, 22]. In the referred studies [19, 21, 22] the effective thermal conductivity (ETC) was higher than that from mono-sized granular beds. Binary mixtures and metallic foams give rise to high ETCs due to the low porosity [19] and low solid-solid contact thermal resistance [21, 22], respectively.

With the help of computational fluid dynamic (CFD) techniques [8, 23-26], in present work we will explore the heat exchange and creeping flow of Newtonian and non-Newtonian liquids in particulate binary mixtures.

2. NUMERICAL SIMULATION

Two dimensional models (figure 2) can be exploited to elucidate complex flow and heat transfer characteristics associated with a porous medium [27]. For mono-size packings (figure 2a), Kuwahara et al. [28] found a reasonable agreement between their numerical results (convective heat transfer coefficients) and the experimental results from Wakao and Kaguei [29]. In the present work we expand the two-dimensional model to binary packings (figure 2b).

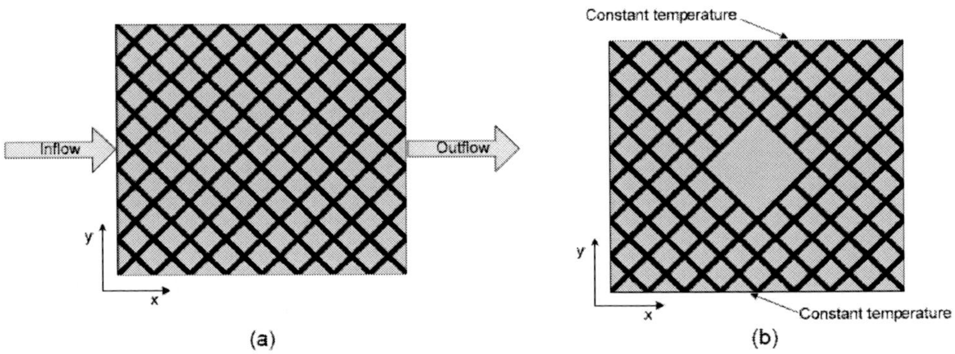

Figure 2. Two of the eight geometries used in the numerical simulations. (a) mono-size packing containing small-sized particles d (packing M), (b) binary packing containing small sized particles d and large particles D (packing B1).

As in previous experimental [19] and numerical [8] works the physical properties from the particles (grey) and fluid (black) are that from glass and water at a temperature of 323.15 K. In the simulations with non-Newtonian fluids (from the power-law type (Eq. (1)) the consistency index, η_0, was that from water at 323.15 K:

$$\eta = \eta_0 \dot{\gamma}^{n-1}. \tag{1}$$

In Eq. (1) $\dot{\gamma}$ represents the shear-rate, n the flow index behaviour and η the apparent viscosity.

The mono-size packing (figure. 2a) had particles (square shape [8, 28]) with size, d, 0.333 mm. The binary packings (figure 2b) contains a mixture of small particles with diameter d and large particles with diameter, D, 1.249 mm. Therefore, the particle size ratio, $\delta = d/D$, was 0.267 in all the simulations. As it will be seen in Section 3, in the present work it was also performed simulations with different values of large particle volume fraction in the mixture, x_D.

The 2D geometries had a length and width of 4.770 mm and 4.165 mm, respectively. The apparent linear flow velocity, temperature imposed in the walls (figure 2b) and water inlet temperature were 3.481×10^{-4} m/s, 353.15 K and 293.15 K, respectively.

The numerical calculations were performed using the commercial finite element software package POLYFLOW®. The equations solved were the conservation of mass, momentum and energy equations for incompressible creeping flow of Newtonian and power-law fluids:

$$\mathrm{div}(\mathbf{u}) = 0, \tag{2}$$

$$\mathrm{div}(\mathbf{T}) + \rho\mathbf{b} - \rho\,\mathrm{div}(\mathbf{uu}) = 0, \tag{3}$$

$$\mathbf{T} \cdot \nabla\mathbf{u} + \rho h - \mathrm{div}(\mathbf{q}) = 0, \tag{4}$$

where $\boldsymbol{q}$ is the heat flux vector, $\boldsymbol{u}$ is the velocity vector, $\boldsymbol{T}$ the total stress tensor and h refers to the heat supply or strength of an internal heater (zero in the present work).

Fourier's law, Eq. (5), governs the heat conduction in the particles,

$$\mathbf{q} = -\lambda_p \nabla T, \tag{5}$$

λ_p being the thermal conductivity of the particles and T the absolute temperature.

The used mesh was constituted by quadrilateral elements [8], figure 3, and the size of the elements (1×10^{-5} m) was fixed after a grid independence test [23-26].

Figure 3. Mesh used in the numerical simulations.

3. Porosity

As already referred, in the present work it were built 2D binary packing geometries (figure 2 and 4) with different values of volume fraction of large particles, x_D. The used values of x_D and porosity, ε, are shown in table 1.

Table 1. Porosity and x_D from the studied geometries.

Packing	x_D	ε (Numerical values)	ε [30]
M	0	0.375	0.375
B1	0.108	0.355	0.361
B2	0.209	0.337	0.340
B3	0.305	0.318	0.321
B4	0.395	0.298	0.306
B5	0.480	0.279	0.296
B6	0.560	0.260	0.291
B7	0.640	0.241	0.290
–	0.7	–	0.294
–	0.8	–	0.307
–	0.9	–	0.333
–	1	–	0.375

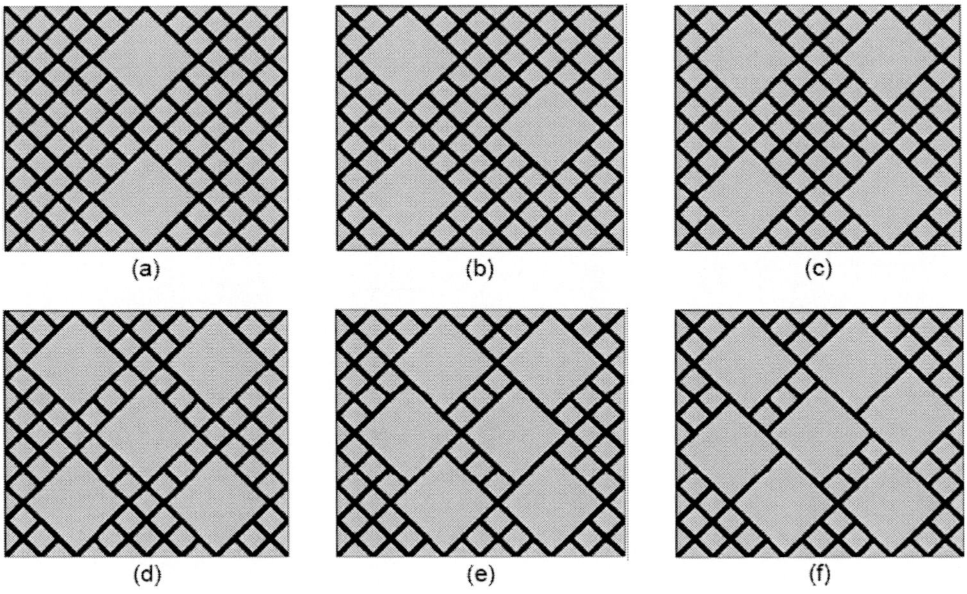

Figure 4. Six of the eight geometries used in the numerical simulations. Packings shown in (a), (b), (c), (d), (d), (e) and (f) are called B2, B3, B4, B5, B6, B7, respectively (see table 1). Packings M and B1 are show in figure 2.

In table 1, packing M and B1 are that shown in figure 2a and figure 2b, respectively. In the same table the porosity obtained numerically is compared with the semi-empirical model from Mota et al. [30], Eqs. (6) to (8). In this model, ε_d and ε_D represents the mono-size packings containing small and large particles, respectively, being assumed $\varepsilon_d = \varepsilon_D = 0.375$, i.e., the porosity from packing M.

The values of porosity from the 2D geometries ($\delta = d/D = 0.267$) are on fairly agreement with the values provided by the referred model, developed using mixtures of glass spheres and for the range $0.1 < \delta \leq 1$.

$$\varepsilon = \left\{1-(1-\varepsilon_D)x_D^{[1.35-x_D F(\delta)]}\right\}\left[\varepsilon_d + (1-\varepsilon_d)x_D^{f(\delta)}\right], \tag{6}$$

$$F(\delta) = -\left\{0.27 - \frac{1.55}{1+1/\exp\left(\dfrac{\delta+0.06}{0.27}\right)}\right\}^{-1}, \tag{7}$$

$$f(\delta) = 5 - 4\sqrt{\delta}. \tag{8}$$

Using the model from Mota et al. [30], in Table 1 it can be also observed that the minimum porosity, ε_{min}, from the glass spheres mixtures is reached for values of x_D close to 0.7. Dias et al. [16] studied the porosity of mixtures of glass spheres with small (≤ 0.1) values of δ and found that the experimental values were well described by the following formula:

$$\varepsilon = \frac{\varepsilon_d(1-x_D)}{1-\varepsilon_d x_D}\underbrace{\exp\left(1.2264x_D^{1/\sqrt{\delta}}\right)}_{\phi}, \tag{9}$$

this equation being applicable in the range $0 < x_D \leq x_{Dmin}$. The value of x_{Dmin} can be calculated by Eq. (10).

$$x_{D\min} = \frac{1-\varepsilon_D}{1-\varepsilon_d\varepsilon_D}. \tag{10}$$

Since ε_d and ε_D assume typical values close to 0.4, this means that x_{Dmin}.is close to 0.7. For values of $\delta \leq 0.1$, minimum porosity, ε_{min}, is reached for $x_D = x_{Dmin}$ and when $\delta \rightarrow 0$ (see Eq. (9)) ε_{min} reaches the absolute minimum porosity, ε_{Min} :

$$\varepsilon_{Min} = \frac{\varepsilon_d(1-x_{D\min})}{1-\varepsilon_d x_{D\min}} = \varepsilon_d\varepsilon_D. \tag{11}$$

For $\varepsilon_d = \varepsilon_D = 0.4$, $\varepsilon_{Min} = 0.16$, i.e., the large particles form a skeleton and the small particles fill the void volume of the referred skeleton (pure filling mechanism [31]). When the particle size ratio differs from $\delta \rightarrow 0$, the mixture is not obtained by a pure filling mechanism since wedging and wall effects disturb the arrangement of large and small particles in the binary mixture and, therefore, ε_{min} is superior to ε_{Min}. In this region ($0 < \delta < 0.1$), small particles affect the large particles by wedging, whereas large particles affect small particles by the wall effect near their own surface [16].

When the particle size ratio assumes moderate values ($0.1 < \delta < 1$), the interaction between the two fractions of different sizes becomes more pronounced [16]. In this case, the packing is controlled by both components (occupation mechanism [31]) since the introduction of a component into the packing results in a dramatic change of the skeleton of the previous component [31].

Using experimental data of minimum porosity ($x_{Dmin} = 0.7$), obtained with different values of δ, Dias et al. [19] found the following relation:

$$\varepsilon_{min} = 0.368 - 0.615 / \{1 + \exp[(\delta + 0.1)/0.153]\}. \tag{12}$$

The main part of the minimum experimental binary packing porosity is located between the bounds [16, 19] defined by Eq. (12) and the model from Liu and Ha [32], applicable for random loose and dense packing, respectively [19]. However, the referred bounds might vary slightly, due to the dependence of ε_d and ε_D on the packing procedure [19].

4. Heat Exchange

An important phenomenon in porous media relates to the enhanced mixing of the local fluid stream due to the tortuous path offered by the complex geometry of a porous medium [33], figure 5. Intensified mixing of the flowing fluid enhances the convection heat transfer [34].

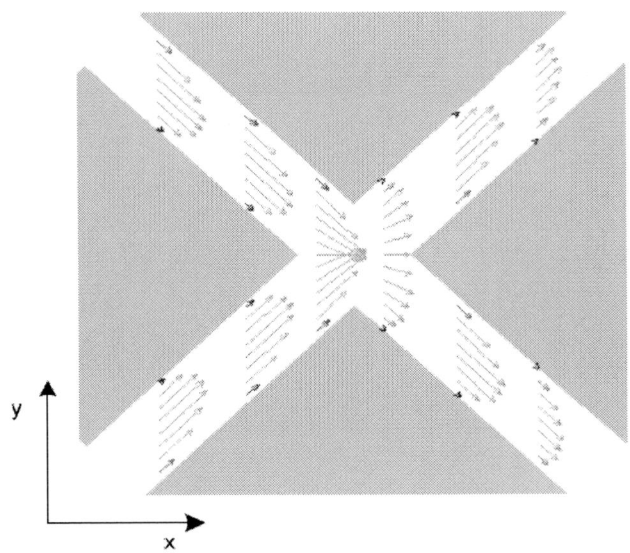

Figure 5. Velocity field in packing M.

The heated packings M, B5 and B7 (figure 6) provide outlet water temperatures 338.98 K, 340.54 K and 342.09 K, respectively. The better thermal performances from the binary packing, when compared with packing M, confirm the experimental results from Dias et al. [19].

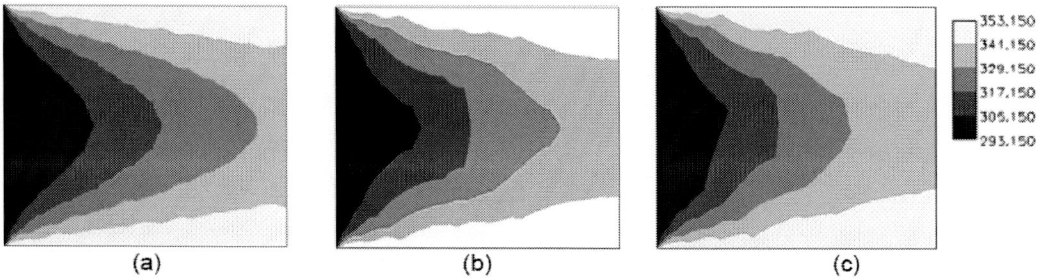

Figure 6. Thermal performance. (a) packing M, (b) packing B5, (c) packing B7.

The improvement of the thermal performance is related to the increase of ETC and transversal thermal dispersion conductivity [19]. In transition from packing M to packings B5 and B7 the porosity decreases and, therefore, ETC (or K_e) increases [35]:

$$\frac{K_e}{K_f} = 1-(1-\varepsilon)^{1/2} + 2\frac{(1-\varepsilon)^{1/2}}{1-\lambda_h U}\left(\frac{(1-\lambda_h)U}{(1-\lambda_h U)^2}\ln\frac{1}{\lambda_h U} - \frac{U+1}{2} - \frac{U-1}{1-\lambda_h U}\right), \tag{13}$$

$\lambda_h = K_f / K_s$ representing the ratio between the fluid, K_f, and solid, K_s, thermal conductivities. U can be calculated by:

$$U = Z\left((1-\varepsilon)/\varepsilon\right)^{10/9}. \tag{14}$$

The value of Z depends on the shape of the particles and Zehner and Schlünder [35] suggested that Z = 1.25 for spheres. Dias et al. [19] found that the ETC from different binary mixtures (porosity varied between 0.395 and 0.198) of glass spheres (saturated with glycerol) was well described by Eq. (13), using Z = 1.25 in Eq. (14).

The model from Zehner and Schlünder [35] is based on the assumption of contact points between particles but it is well known that finite area contacts between spheres occur in a packed-sphere bed [36-42]. A particle-particle fraction contact area, c/a, was used by Nozad et al. [39] on their 2D numerical model (figure 7a), as well as Hsu et al. [37] on their 3D touching cubs analytical model (figure 7b).

The analytical model from Hsu et al. [37] is given by:

$$\frac{K_e}{K_f} = 1-\gamma_a^2 - 2\gamma_a\gamma_c + 2\gamma_c\gamma_a^2 + \frac{\gamma_c^2\gamma_a^2}{\lambda_h} + \frac{\gamma_a^2(1-\gamma_c^2)}{1+\gamma_a(\lambda_h-1)} + \frac{2\gamma_a\gamma_c(1-\gamma_a)}{1+\gamma_a\gamma_c(\lambda_h-1)}, \tag{15}$$

$$1-\varepsilon = (1-3\gamma_c^2)\gamma_a^3 + 3\gamma_c^2\gamma_a^2, \tag{16}$$

γ_c representing the particle-particle fraction contact area (c/a) and assuming a value of 0.13 for a packing of spheres [36-38]. Using the packing porosity, the value of the dimensionless parameter γ_a (a/l in figure 7b) can be estimated. Eq. (15) and the values of γ_c and γ_a can then be used in order to determine the ETC of the porous medium saturated with a fluid.

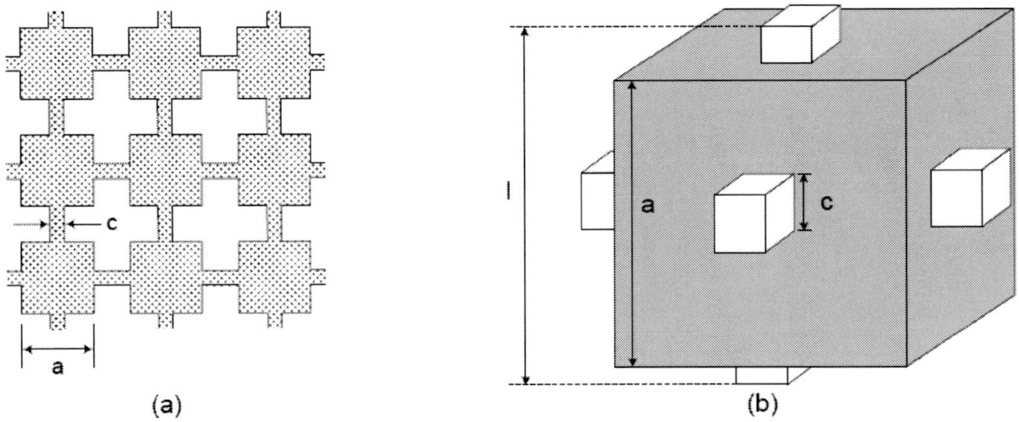

Figure 7. Effective thermall conductivity. (a) unit cell from Nozad et al. [39], (b) unit cell from Hsu et al. [37].

For low (< 10^3) values of $\lambda_h^{-1} = K_s/K_f$ the models from Zehner and Schlünder [35], Nozad et al. [39] and Hsu et al. [37] provide similar values of ETC. When $\lambda_h^{-1} = K_s/K_f >$ 10^3 the solutions found by the models from Nozad et al. [39] and Hsu et al. [37] are similar but the model from Zehner and Schlünder [35] underpredicts the ETC [36].

For values of $K_s/K_f < 10^3$, conduction through the finite contact area loses importance and other mechanisms such as the solid-fluid-solid conduction controls the amount of transferred heat, resulting in a good agreement between experimental data and Zehner and Schlünder model. Solid-solid heat conduction through the finite contact area between spherical particles becomes the main mechanism of heat transfer when the ratio between the solid and fluid conductivities, K_s/K_f, is higher than 10^3 and this seems to be the reason for the deviation of Zehner and Schlünder model from experimental measurements, since is based on the assumption of contact points between particles and not on a finite contact area [19, 36, 41, 42].

As already referred, the better thermal performance of the binary packings (figure 6) is also related to the increase of the transversal thermal dispersion conductivity, K_T. This conductivity may be estimated by the model from Du et al. [43]:

$$\frac{K_T}{K_f} = D_f(\tau - 1)\left(1 - \frac{\varepsilon}{\tau}\right)\mathrm{Re}_p \mathrm{Pr}_f, \qquad (17)$$

Re_p and Pr_f being the Reynolds and Prandtl numbers and D_f an empirical constant assumed to be 0.35 [43]. The tortuosity coefficient, τ, is defined by the ratio between the average travel flow pathway length and the packing bed thickness. The tortuosity coefficient may be also estimated by [43]:

$$\tau = \frac{u_i}{u / \varepsilon}, \tag{18}$$

u_i being the average interstitial velocity and u the apparent linear flow velocity (or Darcy velocity). Since u and ε are known and the average interstitial velocity may be calculated using POLYFLOW® [23, 26], τ may be estimated by Eq. (18). In figure 8 it is shown the normalized tortuosity coefficient [44], τ^*, for each of the geometries presented in Table 1. In figure 8 $\tau*$ is defined by the ratio between τ and the tortuosity coefficient observed in packing M for the Newtonian fluid (n = 1 in Eq. (1)).

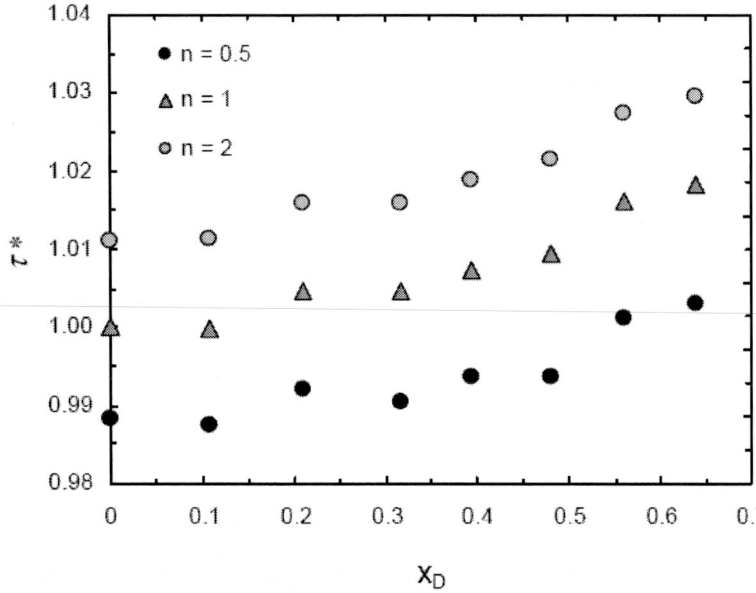

Figure 8. Normalized tortuosity, τ^*, for Newtonian (n = 1) and non-Newtonian fluids (n = 0.5 and 2).

Despite the moderate value of δ (0.267), the numerical results confirm that in the region 0 < x_D < 0.65, the tortuosity coefficient increases with the increase of x_D [18, 44]. In addition, in figure 8 it may be seen that the tortuosity coefficient decreases with the decrease of n (see Eq. 1), due to the decrease of the local interstitial velocities, u_i, figure 9.

Balhoff and Thompson [45] detected the same effect using shear-thinning fluids (n < 1 in Eq. (1)) and mono-size pakings. The decrease of τ with the decrease of n was also reported by Fernandes et al. [26] in a study devoted to the 3D fluid flow of Newtonian and power-law fluids in the complex passages of chevron-type plate heat exchangers.

For Newtonian fluids, Eq. (17) predicts that the tortuosity increase and porosity decrease lead to higher values of K_T. Heat exchange enhancement observed figure 6 can be explained by the increase of ETC, due to the porosity reduction associated to binary mixtures, and by the increase of K_T, due to the tortuosity increase and porosity decrease [19].

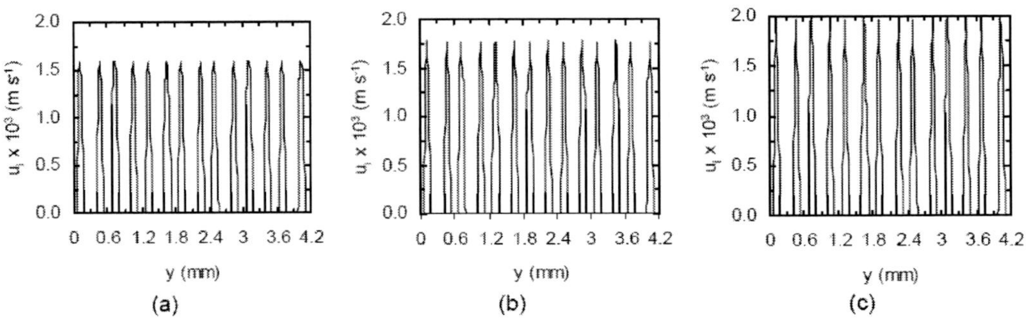

Figure 9. Local interstitial velocities, u_i, in packing M. (a) $n = 0.5$, (b) $n = 1$ (Newtonian fluid) and (c) $n = 2$.

5. PERMEABILITY

Besides K_T, tortuosity and porosity affect packing permeability, k [46-49]:

$$k = \frac{d_{av}^2 \varepsilon^3}{36 K_0 \tau^2 (1-\varepsilon)^2}, \tag{19}$$

where K_0 is a shape factor that depends on the cross-section shape of the capillaries present in granular beds, being $K_0 = 2$ for a packing of spheres [18, 30]. For binary mixtures of large D and small d spheres the average particle diameter is given by [17-19, 30]:

$$d_{av} = \left(\frac{x_D}{D} + \frac{1 - x_D}{d} \right)^{-1}. \tag{20}$$

The tortuosity coefficient depends on the overall porosity, a power-law function being the most frequently used to describe the relation between τ and ε [17-19, 30, 50-52]:

$$\tau = \frac{1}{\varepsilon^{\alpha}}, \tag{21}$$

where α is a coefficient dependent on the packing properties and ranges from 0.4 (loose packing) to 0.5 (dense packing) [17-19, 30, 50-52].

When tiny particles are close to large ones, large-sized particles act as a wall, this effect (wall effect) causing a bypass of a fraction of liquid through the less dense packing near the surface of large-sized particles [16, 18]. Therefore, the real average flow pathway is smaller than the expected for a dense packing ($\alpha = 0.5$).

Dias et al. [18] observed that α assumed a value of 0.5 and 0.4 for $x_D = 0$ and x_{Dmin}, respectively. The decrease of α from 0.5 to 0.4 was explained by the increase of the surface area fraction of large particles, S_D, present in the mixtures and, therefore, by the increase of the fraction of porous media involved in the wall effect, Eq. (22).

$$\alpha = 0.5 - 0.062\left(\delta^{-1} - 1\right)\underbrace{\frac{x_D}{x_D + \left(1 - x_D\right)\delta^{-1}}}_{s_D}. \tag{22}$$

Eq. (22) is applicable in the range $0.03 < \delta \leq 1$, since it is expected that α gradually returns to 0.5 for values of α inferior to 0.03 [17]. When δ reaches a value of 0.0035 the wall effect becomes negligible and therefore the binary packing approaches a regular packing, being expected that α reaches once more the value of 0.5 in the referred δ region [16-18].

Eqs. (12), (21) and (22) may be used to predict the behaviour of τ for different values of δ in the region of ε_{min} ($x_D = x_{Dmin}$), figure 10.

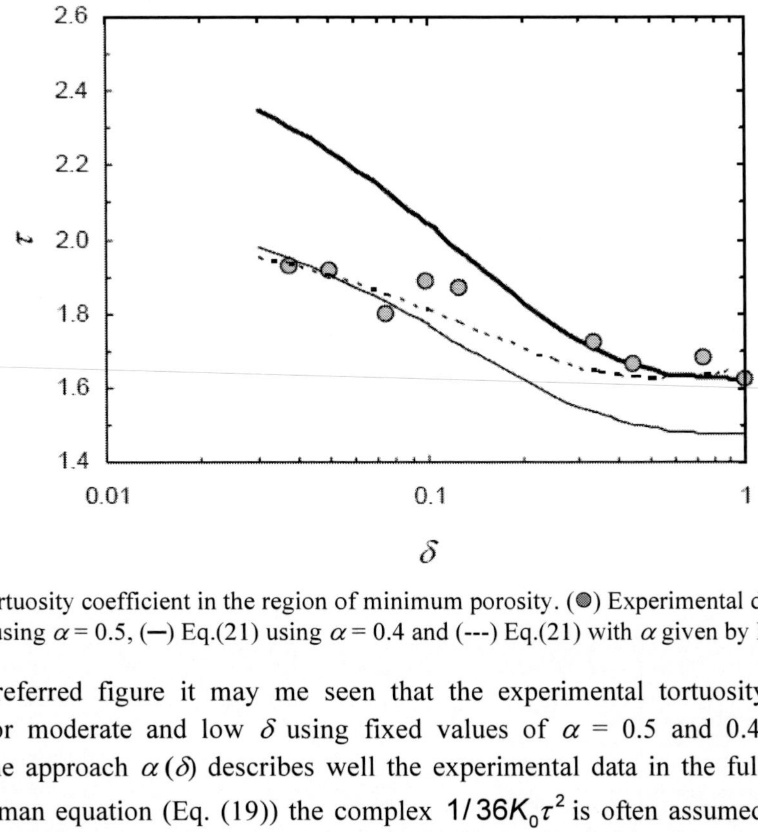

Figure 10. Tortuosity coefficient in the region of minimum porosity. (●) Experimental data [17, 18], (━) Eq.(21) using $\alpha = 0.5$, (—) Eq.(21) using $\alpha = 0.4$ and (---) Eq.(21) with α given by Eq. (22).

In the referred figure it may me seen that the experimental tortuosity data is well described for moderate and low δ using fixed values of α = 0.5 and 0.4, respectively. However, the approach $\alpha(\delta)$ describes well the experimental data in the full range. In the Kozeny-Carman equation (Eq. (19)) the complex $1/36K_0\tau^2$ is often assumed as a constant 1/150 or 1/180. In figure 9, for $\delta = 0.0375$, this complex takes the values 1/274 and 1/382 for $\alpha = 0.4$ an 0.5, respectively, the experimental value being 1/267. Using these different values of $1/36K_0\tau^2$ in Eq. (19), the permeability of a given binary packing may differ about 60 %.

Dias et al. [18, 19] calculated the binary packings permeability using the measured Darcy velocity at a fixed pressure drop. Using the experimental permeability the values of τ and α were calculated [17, 19] using Eqs. (19) and (21), respectively. Using spheres, sand, spheres mixtures, sand mixtures and spheres/sand mixtures, Currie [53] found porosities between 0.424 (sand) and 0.171 (spheres/sand mixture) and tortuosities between 1.549 (α = 0.51 in Eq. (21)) and 2.175 (α = 0.44), respectively. Currie [53] measured the effective diffusion coefficients of hydrogen in dry samples of the above referred materials, the main part of the α values being also located in the approximate range 0.4 – 0.5 [17, 19].

Using Eqs. (19) to (21) at a fractional content x_{Dmin}, correspondent to the region of minimum porosity $\varepsilon_{\min}$, the dependence of the dimensionless ratio k_{bp}/k_d on δ is represented by the expression:

$$\frac{k_{bp}}{k_d} = \left(\frac{1-\varepsilon_d}{1-\varepsilon_{\min}}\right)^2 \left(\frac{\varepsilon_{\min}}{\varepsilon_d}\right)^{3+2\alpha} \left(\frac{1}{\delta x_D + 1 - x_D}\right)^2 . \tag{23}$$

On Eq. (23) k_{bp} stands for binary packing permeability, this packing containing spheres with diameter *D* and spheres with diameter *d*. The permeability of the mono-size packing of spheres with diameter *d* is represented by k_d [19].

Combining Eq. (23) with Eqs. (12) and (22) to describe the minimum porosity and α, respectively, a good agreement with the experimental data is obtained (figure 11). The δ region where the permeability of binary packing, k_{bp}, is higher than the permeability of mono-size packing of particles *d*, k_d, is located in the range $0.1 \le \delta < 1.0$.

In figure 11 it can be also observed that an increase in permeability by a factor of two is achieved for particle size ratios between 0.3 and 0.5. Recently, Dueck [54] noticed a similar effect.

Having in mind the thermal results reported in Section 4 (see figure 6) and figure 11, particulate binary mixtures may give rise to a substantial improvement of the thermal-hydraulic performance associated to mono-size packing [19].

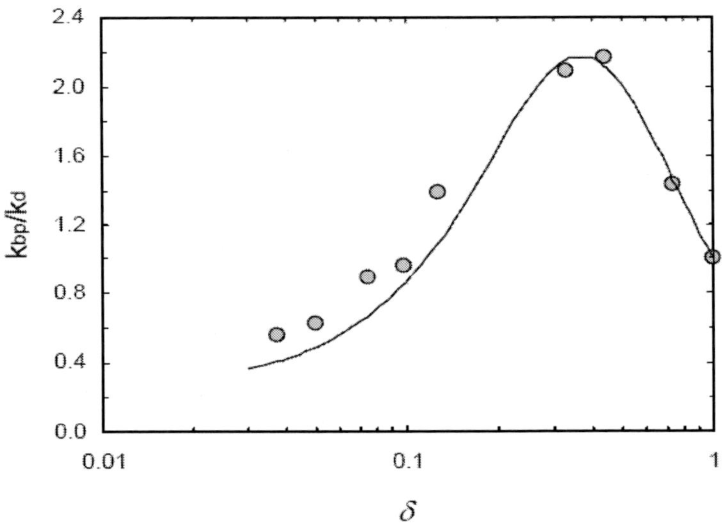

Figure 11. Dependence of k_{bp}/k_d on the particle size ratio in the region of minimum porosity. (●) Experimental data [19], (—) Eq. (23) with α given by Eq. (22) and $\varepsilon_{\min}$ given by Eq. (12).

6. Conclusion

Granular beds composed of different particle sizes have a wide application in industry (trickle-bed reactors, chromatographic columns, solid-liquid separation, heat exchangers, etc.) and science, giving rise to a wide range of values of permeability, porosity, particle size or pore size, tortuosity, effective thermal conductivity, thermal dispersion conductivities, etc. The study of these porous media properties is difficult when the pore topology is difficult to control. This is the case of natural porous media, such as soil, made of irregular multisized particles. A way to overcome this setback is to use controlled packed beds containing mixtures of different sized spherical particles, which may serve as experimental models.

References

[1] Rachwal, A. J.; Bauer, M. J.; Chipps, M. J.; Colbourne, J. S.; Foster, D. M. In *Advances in slow sand and alternative biological filtration*; Graham, N.; Collins, R., eds; John Wiley & Sons: Chichester, 1996, pp. 3-9

[2] Lage, J. L. In *Transport Phenomena in Porous Media;* Ingham, D. B.; Pop, I., eds.; Pergamon: Oxford, 1998, pp. 1-30.

[3] Darcy, H. P. G. *Les Fontaines Publiques de la Ville de Dijon*; Victor Dalmont: Paris, 1856.

[4] Provder, T. *Prog. Org. Coat.* 1997, 32,143-153.

[5] Small, H. *J. Colloid Interface Sci.* 1974, 48,147-161.

[6] Stegeman, G.; Kraak, J. C.; Poppe, H. *J. Chromatogr. A* 1993, 657, 283-303.

[7] Venema, E.; Kraak, J. C.; Poppe, H.; Tijssen, R. *J. Chromatogr. A* 1996, 740,159-167.

[8] Dias, R. P.; Fernandes, C. S.; Mota, M.; Teixeira, J. A.; Yelshin, A. *Carbohydr. Polym.* 2008, doi:10.1016/j.carbpol.2008.05.001.

[9] Dias, R. P. *Recent Patents Eng.* 2008, 2, 95-103

[10] Boyes, B. E.; Walker, D. G.; McGreer, P. L. *Anal. Biochem.* 1988, 170, 127-134.

[11] Hirabayashi, J.; Kasai, K. *Anal. Biochem.* 1989, 178, 336-341.

[12] Hirabayashi, J.; Kasai, K. *J. Chromatogr. A* 1996, 722, 135-142.

[13] Guillaume, Y. C.; Perrin, F. X.; Guinchard, C.; Nicod, L.; Truong, T. T.; Xicluna, A.; Millet J.; Thomassin, M. *Anal. Chem.* 2002, 74, 1217-1222.

[14] Al-Dahhan, M. H.; Wu, Y.; Dudukovic, M. P. *Ind. Eng. Chem. Res.*, 1995, 34, 741-747.

[15] Dias, R. P.; Teixeira, J. A.; Mota M.; Yelshin, A. *Sep. Purif. Techno.* 2004, 37, 69-80.

[16] Dias, R. P.; Teixeira, J. A.; Mota M.; Yelshin, A. *Ind. Eng. Chem. Res.* 2004, 43, 7912-7919.

[17] Dias, R. P.; Teixeira, J. A.; Mota M.; Yelshin, A. *Sep. Purif. Techno.* 2006, 51, 180-184.

[18] Dias, R. P.; Fernandes, C. S.; Teixeira, J. A.; Mota, M.; Yelshin; A. *J. Hydrol.* 2008, 349, 470-474.

[19] Dias, R. P.; Fernandes, C. S.; Teixeira, J. A.; Mota, M.; Yelshin; A. *Int. J. Heat Mass Transfer* 2007, 50, 1295-1301.

[20] Jiang, P.-X.; Fan, M.-H.; Si, G.-S.; Ren, Z.-P. *Int. J. Heat Mass Transfer* 2001, 44, 1039-1051.

[21] Paek, J. W.; Kang, B. H.; Kim, S. Y.; Hyun, J. M. *Int. J. Thermophysics* 2000, 21, 453-464.
[22] Bhattacharya, A.; Calmidi, V. V.; Mahajan, R. L. *Int. J. Heat Mass Transfer* 2001, 45, 1017-1031.
[23] Fernandes, C. S.; Dias, R. P.; Nóbrega, J. M.; Maia, J. M. *Chem. Eng. Process*. 2007, 46, 825-833.
[24] Fernandes, C. S.; Dias, R. P.; Nóbrega, J. M.; Afonso, I. M.; Melo, L. F.; Maia, J. M. *J. Food Eng.* 2005, 69, 281-290.
[25] Fernandes, C. S.; Dias, R. P.; Nóbrega, J. M.; Afonso, I. M.; Melo, L. F.; Maia, J. M. *J. Food Eng.* 2006, 76, 433-439.
[26] Fernandes, C. S.; Dias, R. P.; Nóbrega, J. M.; Maia, J. M. *J. Food Eng.* 2008, doi: 10.1016/j.jfoodeng.2008.05.022.
[27] Nakayama, A.; Kuwahara, F. In *Handbook of porous media*; Vafai, K. eds; Marcel Dekker: New York, 2000, pp. 441-488.
[28] Kuwahara, F.; Shirota, M.; Nakayama, A. *Int. J. Heat Mass Transfer* 2001, 44, 1151-1159.
[29] Wakao, N.; Kaguei, S. *Heat and Mass Transfer in Packed Beds*; Gordon Breach Science Publishers: New York, 1982, pp. 243-295.
[30] Mota, M.; Teixeira, J. A.; Yelshin, A. *Trans. Filtration Soc.* 2001, 1, 101–106.
[31] Yu, A. B.; Standish, N. *Ind. Eng. Chem. Res.* 1991, 30, 1372-1385.
[32] Liu, S.; Ha, Z. *Powder Techno.* 2002, 126, 283-296.
[33] Vafai, K.; Amiri, A. In *Transport Phenomena in Porous Media;* Ingham, D. B.; Pop, I., eds.; Pergamon: Oxford, 1998, pp. 313-329.
[34] Jiang, P. X.; Wang, Z.; Ren, Z. P.; Wang, B. X. *Exp. Thermal Fluid Sci.* 1999, 20, 45-54.
[35] Zehner, P.; Schlünder, E.U. *Chemie-Ingenienr-Technik.* 1970, 42, 933-941.
[36] Cheng, P.; Hsu, C. T. In *Transport Phenomena in Porous Media;* Ingham, D. B.; Pop, I., eds.; Pergamon: Oxford, 1998, pp. 57-76.
[37] Hsu, C. T.; Cheng, P.; Wong, K.W. *J. Heat Transf.* 1995, 117, 264-269.
[38] Hsu, C. T. In *Handbook of porous media*; Vafai, K. eds; Marcel Dekker: New York, 2000, pp.170-200.
[39] Nozad, I.; Carbonell, R. G.; Whitaker, S. *Chem. Eng. Sci.* 1985, 40, 843-855.
[40] Nozad, I.; Carbonell, R. G.; Whitaker, S. *Chem. Eng. Sci.* 1985, 40, 857-863.
[41] Argento, C.; Bouvard, D. *Int. J. Heat Mass Transfer* 1996, 39, 1343-1350.
[42] Cheng, G. J.; Yu, A. B.;Zulli, P. *Chem. Eng. Sci* 1999, 54, 4199-4209.
[43] Du, J.; Hu, X.; Wu, W.; Wang, B.-X. Heat Transfer - Asian Research 2003, 32, 545-552.
[44] Mota, M.; Teixeira, J. A.; Yelshin, A. *Sep. Pur. Techno.* 1999, 15, 59-68.
[45] Balhoff, M. T.; Thompson, K. E. *Chem. Eng. Sci.* 2006, 61, 698-719.
[46] Carman, P. C *Trans IChemE* 1937, 15, 150-166.
[47] Bear, J., 1972. Dynamics of Fluids in Porous Media. Dover Publications, Inc., New York, pp. 112–167.
[48] MacDonald, M. J.; Chu, C.-F.; Guilloit, P. P.; Ng, K. M. *AIChE J.* 1991, 37, 1583–1588.
[49] Ben Aim, R.; Le Goff, P.; Le Lec, P. *Powder Techno.* 1971, 5, 51-60.
[50] Millington, R. J.; Quirk, J. P. *Trans. Faraday Soc.* 1961, 57, 1200–1207.

[51] Klusácek, K.; Schneider, P. *Chem. Eng. Sci.* 1981, 36, 523–527.
[52] Zhang, T. C.; Bishop, P. L. *Water Res.* 1994, 28, 2279–2287.
[53] Currie, J. A. *J. Appl. Phys.* 1960, 11, 318-323.
[54] Dueck, J. G. J. *Eng. Phys. Thermophys.* 2007, 80, 662-670.

In: Porous Media: Heat and Mass Transfer…
Editors: J. L. Acosta and A.F. Camacho
ISBN: 978-1-60692-437-2

Chapter 8

PARTICLE TRANSPORT AND DEPOSITION IN POROUS MEDIA

Ahmed Benamar, Nasre Dine Ahfir and Hua Qing Wang

ABSTRACT

The transport mechanisms of colloids in saturated porous media have been studied in great details. Suspended particle (SP) transport in the subsurface has only recently attracted significant attention. This chapter presents an experimental study of the transport of suspended particles in a saturated porous media, aimed at delineating the effects of hydrodynamic and gravity forces on particle transport and deposition rate. Suspended particles were injected under saturated flow conditions into a laboratory column packed with gravel or glass beads. The measured particle breakthrough curves were well described by the analytical solution of an advection-dispersion equation with a first-order deposition kinetic. The laboratory tests performed with different flow rates and column materials showed that hydro-dispersive parameters derived from 1D advection-dispersion model depend on the flow rate, the suspended particles characteristics and the porous medium. The results provided the existence of a flow velocity beyond which particles travel faster than the conservative tracer. Hydraulically equivalent media with neighbouring porosities produce differences in the transport of suspended particles. The dispersion was found to be controlled by the pore-space geometry, the relative size of particles and the flow rate. This chapter shows that particles, pore sizes and flow rate are among the main mechanisms influencing the transport and deposition rate of SP in porous media.

Key words: Porous media; transport; Suspended Particles; dispersion; deposition; ionic strength.

1. INTRODUCTION

The colloid transport in saturated porous media has been studied in great details. Owing to their large size and density, suspended particle transport has been only recently attracted significant attention (Frey *et al*, 1999; Gohr Pinheiro *et al*, 1999; Wang *et al*, 2000; Massei *et al*, 2002; Benamar *et al*, 2005; Ahfir *et al*., 2008). Suspended particles transport interests many application fields. Among them, let us point out the industrial fields concerned by various techniques of separation and filtration (chromatography, ultra filtration, water treatment ...). It is necessary to also quote the many fields of "Earth Science" where SP play a major role: Hydrogeology, geo-environment, earth structures safety. During major floods, erosion of alluvial soil causes the occurrence of significant turbidity. Turbid water travels through the alluvium and can reach water supply. The alluvial aquifer might be of interest for the understanding the water pollution risk. Mobile particles in groundwater aquifers and soils can serve as carriers for contaminants and thereby facilitate contaminant transport (Ryan *et al*, 1996). Particle transport and deposition are also the basis of deep-bed (granular) filtration. They also contribute to develop internal erosion in dams and dykes. The presence of large hydraulic gradients combined with clay soils that may be inadequately filtered creates the potential for internal erosion and possible hydraulic failure of the dyke. A thorough understanding of particle filtration is essential for predicting the transport and fate of microbial particles, such as bacteria and colloid-bound pollutants in subsurface environments. Predicting the importance of particle-facilitated transport is difficult because the processes governing particle release, transport, and deposition in natural subsurface environments are not yet fully understood. Experiments in an undisturbed, saturated column of porous medium using fluorescent latex microspheres as tracers indicate that particle diameter plays a major role in controlling transport (Cumbie and McKay, 1999). Major studies on colloid deposition in porous media were firstly performed using step-input experiments. More recently, short-pulse breakthrough experiments have been used to study colloid and particle deposition kinetics in model systems and soils (Kretzschmar et al., 1997; Compere et al., 2001; Massei et al., 2002). Short pulse experiments have the advantage that much smaller amounts of particles are introduced into the column, thereby minimizing possible effects of retained particles on deposition rate.

Various restrictions on the movement of suspended particles such as clogging, mechanical straining and filtration were studied. During the flow of the suspension through the medium, particle transport and capture result from several forces and mechanisms depending on particle size, pore distribution and flow rate. For larger particles, those typically higher than 10μm, hydrodynamics, gravity and inertial effects are dominant, while all forces and mechanisms can contribute for smaller particles whose size is ranging between 0.1 and 10μm (Gohr Pinheiro *et al*, 1999). The basis for understanding flow and transport phenomena in porous media has largely been developed from experimental and theoretical studies in macroscopic systems in which coupled behaviour at the pore scale is not measured or observed directly. Experimentation has focused on measuring average medium properties or process variables, such as the porosity, permeability, fluid pressure or particles concentration, in laboratory column. The ability to predict and quantify contaminant migration in soils is essential for risk assessment.

This chapter is dealing with the behaviour of suspended particles in porous media and their role in contaminant transport and erosion in earth structures. It addresses transport and deposition of suspended particles in two porous media that are different in pore geometry under water-saturated flow conditions. This chapter investigates the various mechanisms governing the process of particle transport, with emphasis on the role of particle deposition.

Particle deposition rates and attachment efficiencies can be determined from step-input or short-pulse column experiments.

2. THEORETICAL BACKGROUND

Under steady state saturated flow conditions, the transport of suspended particles through granular porous media can be described by a convective-dispersive transport equation including terms to account for particle deposition and release (de Marsily, 1986):

$$\frac{\partial C}{\partial t} = D_L \frac{\partial^2 C}{\partial x^2} - u\frac{\partial C}{\partial x} - \frac{\rho}{w}\frac{\partial S}{\partial t} \tag{1}$$

$$\frac{\rho}{w}\frac{\partial S}{\partial t} = k_d C - \frac{\rho}{\omega} k_r S \tag{2}$$

These equations describes the evolution of the particle concentration in suspension C(x,t) and the amount of deposited particles per unit mass of the porous matrix S(x,t) as a function of travel distance x and time t. Here, D_L is the longitudinal hydrodynamic dispersion of suspended particles, u the average interstitial velocity of particles, ρ the solid matrix bulk density, ω the porosity, and k_d and k_r the particle deposition and release rate coefficients, respectively. The initial kinetics of particle deposition and release are assumed to follow a pseudo-first-order kinetic rate law. This type of equation has been usually used to describe colloid or bacteria transport in laboratory scale columns. Under certain conditions (chemistry, amount of injected particles), particle release rates are very small. The release term can then be dropped from Eq. (2) and the transport equation reduces to:

$$\frac{\partial C}{\partial t} = D_L \frac{\partial^2 C}{\partial x^2} - u\frac{\partial C}{\partial x} - k_d C \tag{3}$$

The above equation cannot be applied when retained particles influence the rate of further particle deposition, as often observed in step-input particle breakthrough experiments in which large amounts of particles are introduced into the porous medium. Equation (3) assumes particle deposition to follow first-order kinetics and to be irreversible. This assumption is justified at sufficiently low particle concentration (i.e., no blocking or ripening) and for moderate to high ionic strengths where particle release is negligible compared to particle deposition (Elimelech et al. 1995; Kretzschmar et al. 1997).

For the short-pulse experiments, the boundary conditions for a semi-infinite medium are given by equation (4).

$$\left.\begin{aligned} &C(t=0,x)=0\\ &C(t,x=0)=m/\ Q\delta(t)\\ &C(t,x=\infty)=0 \end{aligned}\right\} \tag{4}$$

where $\delta(t)$ is Dirac function, m is the masse of particles injected [M], and Q is the flow rate [$L^3.T^{-1}$]. The last boundary condition ($C(t,x=\infty)=0$) is not realistic for finite column experiments, but necessary for providing an analytical solution. The analytical solution for equation (3) according to equation (4) is given by (Wang et al., 2000):

$$C(t,x)=\frac{mx}{Q\sqrt{4\pi D_L t^3}}\exp(-K_d t)\exp\left(-\frac{(x-ut)^2}{4D_L t}\right) \quad (5)$$

Based on the analytical solution (Eq. 5), the regression parabolic method (Wang 2001) is used in order to interpret the breakthrough data. Regression parabolic method (RPM) is a derivative as an extension of the linear graphical method (Wang et al. 1987) to deal with problems involving three parameters. In our tracer tests this method allows to determine at once: the dispersion coefficient D_L, the convection time t_c, and the deposition rate coefficient K_d.

The above model based on the advection-dispersion equation coupled with first-order deposition kinetics was used to simulate the experimental breakthrough curves and to estimate kinetic coefficients describing particle retention in porous media. This simplified model enables an easier fitting because of its limited set of parameters.

3. Materials and Experimental Procedure

A series of well-controlled experiments have been performed in order to address a number of parameters governing suspended particle transport. The experiments were aimed at determining the combined effect of flow velocity and porous media structure. The packed-bed column system used in the experiments is depicted in figure 1. The column consisted of a plexi-glass cylinder with inner diameter of 89 mm. The cylinder was filled with quartz gravel or glass beads (see table 1).. In order to obtain a homogeneously packed porous medium, the column was partially filled with water and subsequent layers (thickness $\approx$ 5 cm) were poured into the column and each layer was packed by vibrating the column. The materials used as porous media are crushed flint gravel and glass beads. The gravel grains distribution was 20% of 1.00 – 2.15 mm and 80% of 2.15 – 3.15 mm; and the glass beads are 3 mm of diameter. Both materials have a uniformity coefficient $C_u < 2$, and so considered uniform. The gravel was treated to remove impurities from the grains. The pore distribution (figure 2) of both materials was measured using mercury porosimeter, showing a mean pore size of 0,73 mm for the glass beads and two modes for the gravel 0,55 mm and 0,73 mm). This pore analysis show the existence of a wide range of pore sizes in the gravel. The column is fed by a reservoir containing water (pH of 6.8 ± 0.1) using a MasterFlex peristaltic pump. A digital flow-meter is installed at the column inlet in order to measure the flow rate. A 60-ml syringe is used to perform pulse injections. The column was equipped with three piezometers allowing the control of hydraulic gradient variations (owing to the particle deposition) during particle transport in porous medium.

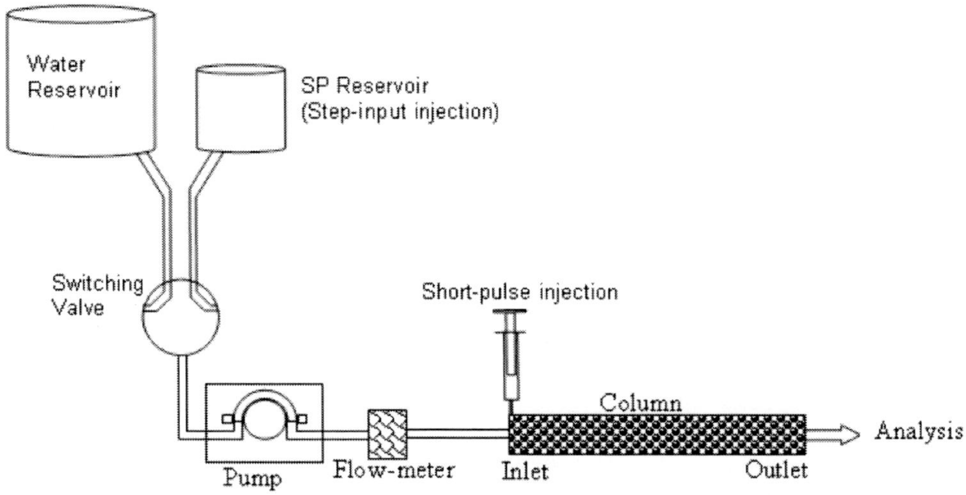

Figure 1. Sketch of the experimental setup of particle transport in porous medium.

Table 1. Selected properties of the two materials used for columns experiments

Porous medium Porosity Hydraulic conductivity Bulk density (cms^{-1})
Crushed gravel 0,47 2,3 1,37
Glass beads 0,42 2,1 1,49

Many experiments were conducted with different flow rates ranging from 0,013 cms^{-1} to 0,437 cms^{-1} under steady state, saturated flow conditions. In each fresh packed column, ten short-pulse injections were performed starting from the highest flow rate to the weakest one. The latter technique is chosen in order to avoid a likely release of the suspended particles previously deposited in the porous medium i.e., the second-order deposition kinetics can be neglected. After each test series the column was repacked with the same material which was previously washed to remove any suspended particles and dried. To compare suspended particles and fluorescein transport behaviour, the injected pulse consists of a mixture of 100 mgl^{-1} of suspended particles and 500 ppb of fluorescein.

The suspended particles are quartz silt collected from surface formations in Haute Normandie (France) and their size ranges from 2 µm to 30 µm with a 14 µm mode (figure 3). The injected pulse is chosen very small (2% of the pore volume) in order to not disturb the water flow in the column even at the lowest rate. The particle tracer was prepared by mixing certain amounts of fines particles (which were previously selected) with water at controlled concentration. The concentration of particles in water samples can be determined in several different ways. In monitoring of groundwater quality, turbidity meters based on optical density measurements (light scattering) are commonly used. Such measurements may strongly depend on particle size and composition. Nevertheless, turbidity is a useful parameter for concentration characterization. The detection system consists of a fluorometer and a turbidimeter which is the nephelometer, determining the turbidity level by measuring the amount of light scattered at 90° by the suspended particles.

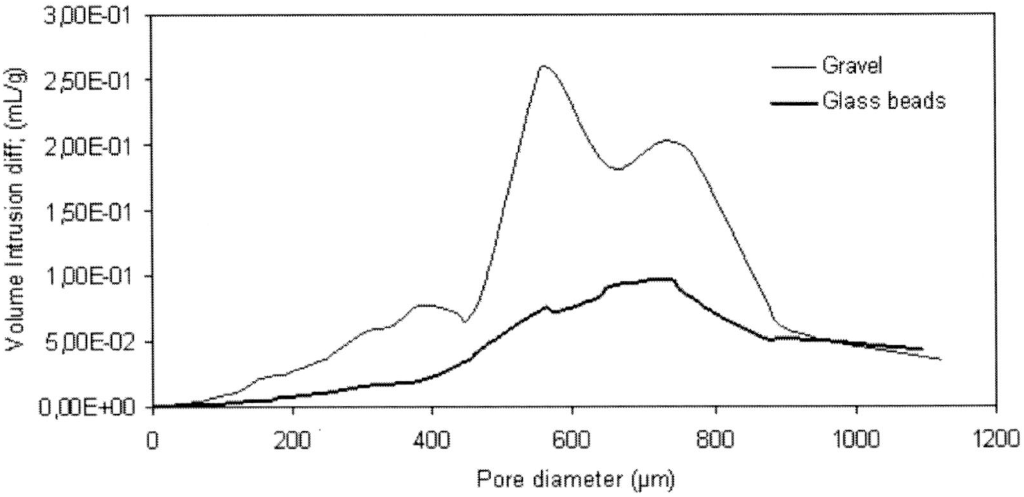

Figure 2. Pore size distribution from mercury porosimeter of the media in-filling the columns.

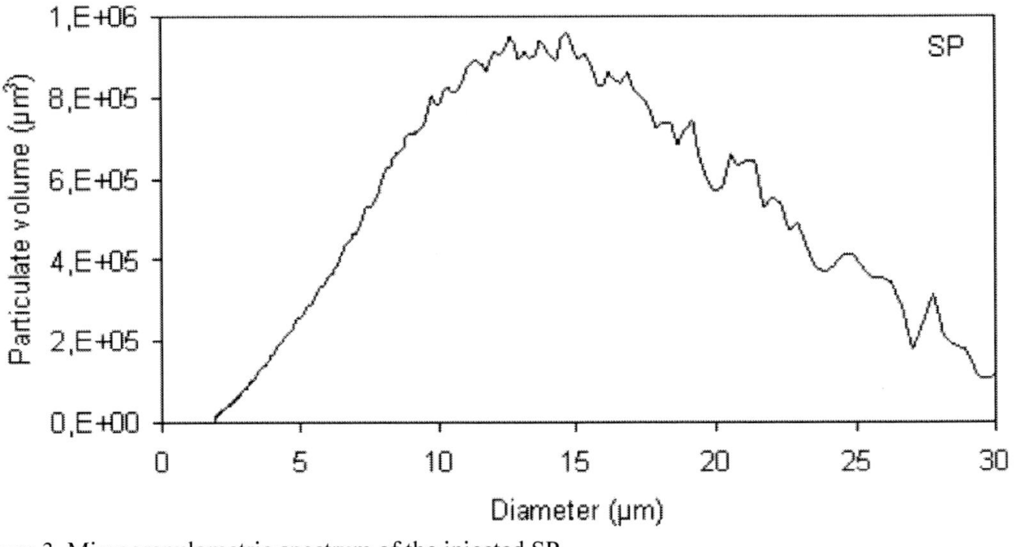

Figure 3. Microgranulometric spectrum of the injected SP.

4. Transfer Process and Size Exclusion Effect

Owing to high velocities tested (mean pore water velocities ranged from 0.03 to 0.95 cm/s), low ionic strength (2.6 mM NaCl), large size of suspended particles (mean size of 14 µm), and neutral water pH, chemical mechanisms may be neglected. In laboratory column studies on the transport of suspended particles through natural porous media it is frequently observed that particles travel considerably faster (up to 50 %) than conservative solute tracers (Massei et al., 2002). This phenomenon is termed the *size exclusion effect*, a well-known effect in chromatography. The size exclusion effect can be explained by exclusion of particles from small pores, leading to a smaller effective pore volume available for particle transport as compared to solute transport. The breakthrough behaviour of the suspended particles in the two porous media investigated governs the particle transport and deposition patterns. The results showed on figure 4 demonstrate the existence of exclusion effect in the two porous media tested. Further results show the existence of a critical flow rate beyond which particles travels faster than the conservative tracer (Ahfir et al, 2007). Flow rate increasing generates the earlier breakthrough of the SP with comparison to that of dissolved tracer. The exclusion of particles from matrix diffusion is the crucial difference between solute and particle tracer transport. The size exclusion effect explains the observed behaviour (Grolimund et al. 1998; Kretzschmar et al. 1997). This behaviour is explained by the existence of preferential pathways through the pore space for different particle sizes (Sirivithayapakorn et al., 2003; Massei et al., 2002). Macro pores in the gravel medium favour faster movement of suspended particles. Size exclusion in gravel leads to a larger-scale pore exclusion effect, where the larger particles are directed through only certain regions of the porous medium (macro pores). The preferential paths become more important for greater flow rates. The earlier breakthrough of particles in comparison with conservative tracer occurs in the glass beads column at a lower velocity than that measured in the gravel column. This behaviour can be explained by analysing the granular size distribution of the effluent (figure 5). This figure displays the changes in the particle size distribution during recovery, analysed with a Coulter Multisizer counter. For the greatest flow rate in the gravel column, the recovery of coarser particles is more important than that of the glass beads column. Thus, for the same flow rate the largest diameter mode of the recovered particles is 14 μm for the gravel and 8 μm for the glass beads (figure 5). This result is also related to the pore structure of the gravel (figure 2) which leads to the existence of macro pore continuity and its importance on preferential transport (Allaire et al., 2002). The studies conducted on a two-dimensional model demonstrate the role played by high permeability zones in particle transport and deposition. This behaviour, named size-exclusion effect (preferential pathways through the pore space for different particle sizes) is the basis of hydrodynamic chromatography process where the larger particles can not be transported close to the wall pore (tube) because of their dimension. Thus they are only transported by higher velocities near the pore axis, whereas the dissolved tracer can approach the wall (lower velocities).

The particle deposition rate is initially high and almost no particles are detected in the effluent after the first pore volume. However, the reduced particle concentration in the effluent slowly increases and finally reaches a value close to unity after about ten pore volumes, meaning that the particle deposition rate has dropped to zero. When the influent is switched to water without particles, the particle concentration in the effluent drops quickly to zero without exhibiting significant tailing, a behaviour which can only be explained by irreversible deposition of the suspended particles. Thus, the matrix surfaces in the porous medium.

Within a cylinder tube, and on the laminar flow regime, the microscopic velocity profile is parabolic, and the maximum velocity occurs at the tube axis. Due to the transversal molecular diffusion, dissolved tracer is transported by all existing velocities in the tube. Thus, it is transported on the total section of the tube. Due to the size of the particles, larger particles can not be transported close to the wall because of their dimension. Theoretically, the ratio of particulate and dissolved tracer velocities does not exceed 1.5 (de Marsily 1986). Within the framework of this chapter the greatest value of this ratio is obtained at high velocity in the shortest column (33 cm length) and is close to 1.18.

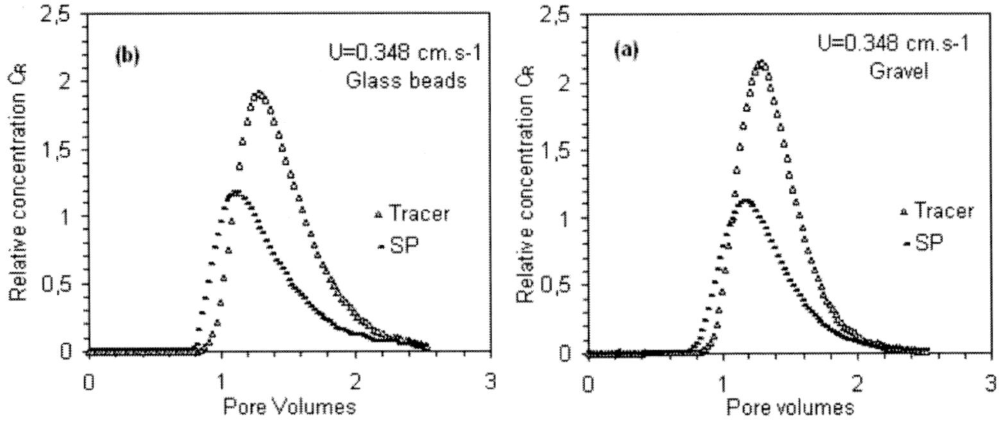

Figure 4. Breakthrough curves showing size exclusion effect in gravel (a) and glass beads (b).

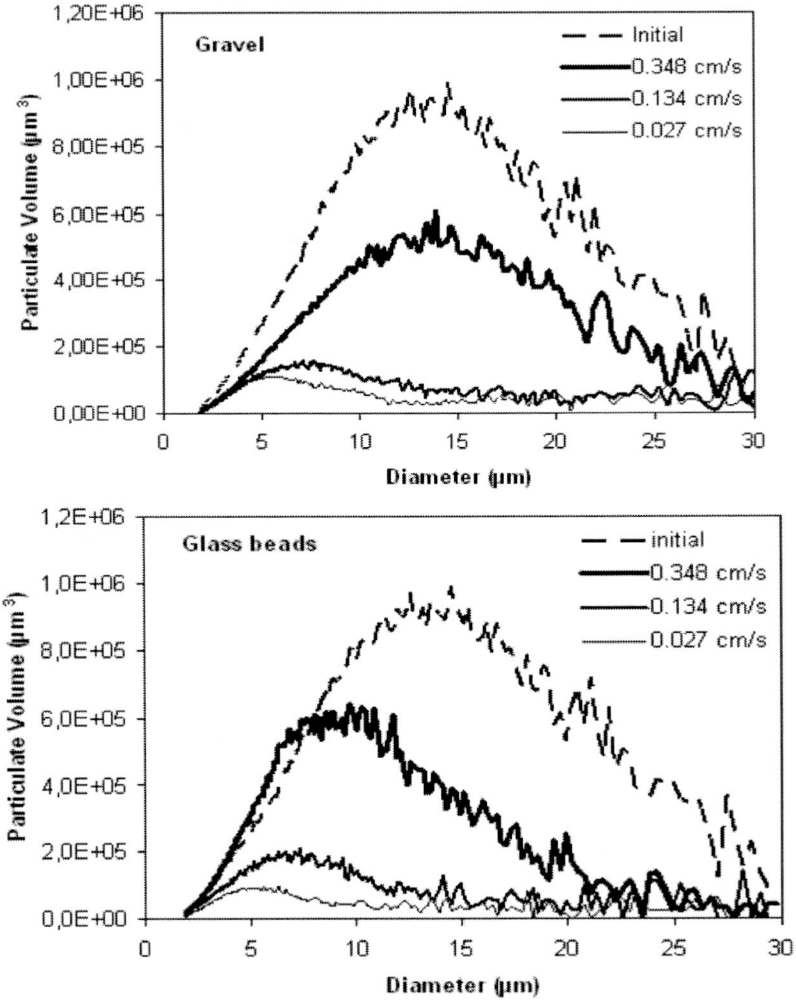

Figure 5. Particle size distribution of the effluent at various flow velocities.

5. Particle Dispersion

Hydrodynamic dispersion represents the spreading of substances in porous media during fluid flow. Under saturated conditions the dispersion coefficient is expressed as:

$$D_L = D_0 + \alpha u^n \quad (6)$$

where D_0 is the effective molecular diffusion of the solute in the porous media (L^2 T^{-1}); α is the dispersivity, and n is an experimental constant. The second term of Eq. 6 is the mechanical dispersion coefficient which is considered, at high Peclet number, to be a linear function of the mean pore water velocity u and which is commonly considered to be a soil characteristic. Typical results suggest that dispersivity is proportional to the grain diameter. Dispersion processes are sensitive to the structure of the medium in which the transport and mixing processes take place. Dispersion in a single capillary tube of constant or varying cross section behaves very differently from that in a disordered porous medium. In a disordered porous medium, the variation in the orientations of flow passages and the coordination of the junctions, which result in wide variations in the length of the streamlines, together with the variations in the geometry of pores and the local pressure gradients, force a chaotic nature on the pore-level velocity field of the flowing fluid. Particle dispersion in porous media is a consequence of the different paths and velocities experienced by the particles. The magnitude of the dispersion at any given flow rate was found to be controlled by the pore-space geometry and the relative size of particles with regards to pore channels (Auset and Keller, 2004). Dispersion coefficient decreases with increasing particle size. Dispersivity is thus not just a function of pore geometry but depends on particle characteristics. Because of their size, larger particles travel in the centre streamlines, leading to faster velocities, less detours, and thus lower range of transit time.

Particle dispersion can in part be explained by pore size exclusion. Exclusion refers to the fact that although soluble tracers are sufficiently small to move into many or all of the pore spaces in the porous media, particles, due to their physical size, may not enter small pores. Sirivithayapakorn and Keller (2003) determined a pore throat to particle diameter threshold of about 1,5 for entering a pore. Larger particles will thus travel through a reduced number of pathways, which on average reduce their travel time.

The dispersivity of homogeneous soil in small-scale column experiments, was usually similar to the mean particle size, while in the field, the dispersivity is many orders of magnitude greater. In the problem investigated here, the molecular Peclet number (ud_{50}/D_0, where d_{50} is the mean particle size) values were found to be between 50 and 100 and hence mechanical dispersion was far more dominant than molecular diffusion.

Using the measured breakthrough curves, dispersion coefficient is derived from Eq. 5. Figure 6 shows the variation of the longitudinal dispersion coefficient as a function of molecular Peclet number for dissolved tracer experiments. A least squares curve fitting yields the relation:

$$\frac{D_L}{D_0} = aPe^m \text{ for } 100 < \mathrm{Pe} < 4000 \quad (7)$$

A similar empirical expression was proposed in literature (Bear, 1972), where a ≈ 0.5 and 1 < m < 1.2 in the range 6 < Pe < 200. The latter is considered to be valid when the main spreading mechanism is caused by mechanical dispersion and molecular diffusion. In the range 200 < Pe < 10^4, the mechanical dispersion is dominant, and the relation between D_L and Pe is assumed to be linear. However, when the experimental data are fitted in the range 100 < Pe < 3500, we still find some nonlinearity. This relation indicates that the dispersivity α_L is not really a constant but increases with flow velocity (figure 7).

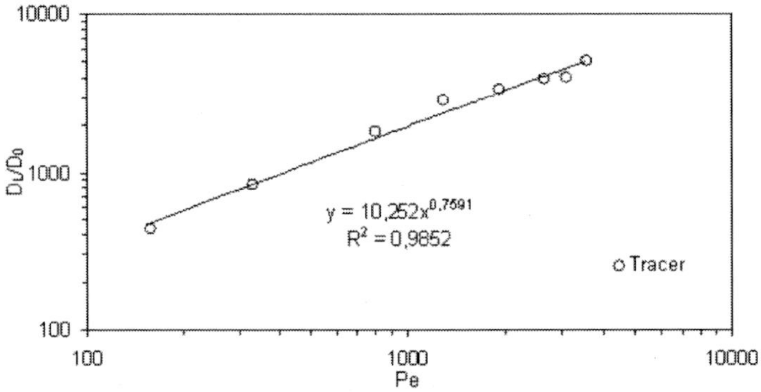

Figure 6. The ratio of the dispersion coefficient and the molecular diffusion coefficient as a function of the molecular Peclet number for the dissolved tracer.

The longitudinal dispersion values obtained from the gravel columns are larger than those obtained from the glass beads columns. Because of the large gravel grains distribution, the pore velocities fluctuations are more important owing to the large pore size distribution. On the other hand, the single-size of the glass beads led to a one-mode pore distribution; thus, one presumes that pore velocities fluctuations are less important than in the gravel medium which thus can be considered here more dispersive than the glass beads medium.

The results further demonstrate that the dissolved tracer dispersion is smaller than that of the suspended particles over all columns investigated (figure 7). This behaviour can be explained by the large size distribution of the suspended particles injected which leads to a large dispersion of the suspended particle velocities.

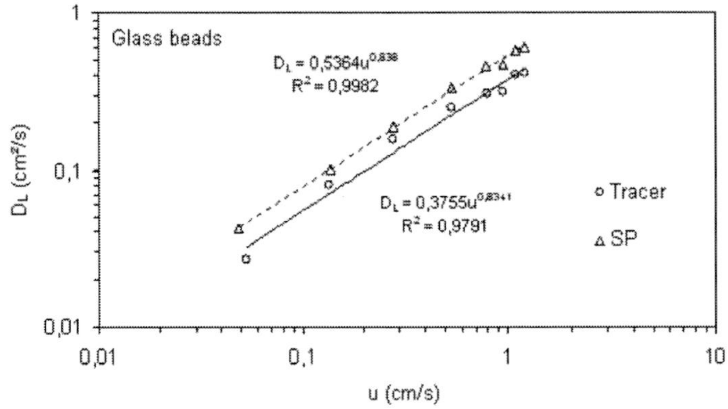

Figure 7. Longitudinal dispersion coefficient as a function of flow velocity for dissolved tracer and suspended particles.

6. Particle Deposition Kinetics - Modelling of Deposition Function

Particle deposition during flow through porous media is commonly assumed to take place in two rate-limiting steps: (1) transport of suspended particles to matrix surfaces (resulting in particle-matrix collisions) and (2) attachment of suspended particles to the matrix surfaces. The fraction of particle-matrix collisions that results in attachment of the suspended particles is called the collision efficiency. The kinetics of the transport step depends on physical factors such as size and density of suspended particles, accessible surface area for particle deposition, pore structure, and flow velocity. The kinetics of the attachment step depends on solution and surface chemistry (interparticle forces between suspended particles and matrix surfaces).

According to DLVO theory, the distance dependent total interaction potential between a suspended particle and a surface can be calculated as the sum of interparticle forces such as Van der Waals attraction and electrostatic repulsion. Particle deposition rates and attachment efficiencies can be determined from step-input or short-pulse column experiments. Few studies have been published on the kinetics of particle deposition in natural porous media, such as soils or aquifers. It has been recognised that the initial removal of particles from suspension during flow through granular porous media follows a pseudo-first-order kinetics rate law, leading to an exponential decrease in suspended particles concentration with travel distance:

$$\frac{C(x)}{C_0} = \exp(-\lambda x) \tag{9}$$

where C_0 is the particle concentration in the column inlet and λ is the filter coefficient. The filter coefficient can be expressed in terms of the macroscopic deposition rate coefficient, k_d, by:

$$\lambda = \frac{k_d}{u} \tag{10}$$

In studies with packed soil columns it has been commonly observed that the particle concentration in the pore water decreases exponentially with travel distance, indicating that particle deposition follows a first-order kinetic rate law in natural porous media. Figure 8 shows the decrease of particle concentration along the porous medium, indicating an exponential trend which characterizes the first order kinetic rate model.

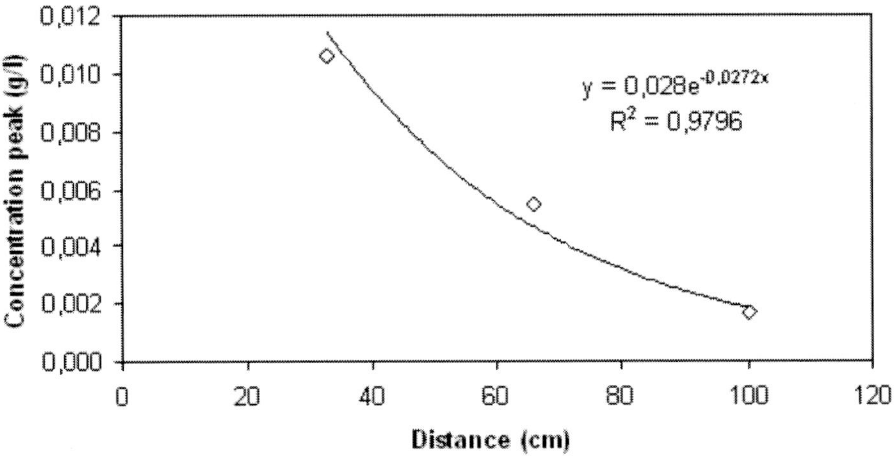

Figure 8. Decrease of particle concentration along the porous medium.

The particle deposition rate as a function of flow velocity for glass beads and gravel column is depicted in Figure 9. The particle deposition rate increases with increasing flow rate until a critical velocity beyond the deposition rate decreases. Usually, for moderate velocities, the deposition rate is described by a power law. Masséi et al. (2002) found a power value of 0.7 for silt particles. Kretzschmar et al. (1997) found a power value of 0.31 for carboxyl latex colloids and 0.18 for Humic-coated hematite colloids. In our case, below the critical velocity, the deposition rate is well described by a power law such as a power value of 0.67 and 0.62 are found within the gravel and glass beads column, respectively. For particles smaller than about 1 µm, where particle deposition is controlled by convection-diffusion, Song and Elimelech (1993) predicted slope to the power of one-third for deposition onto impermeable solid surface, such as spherical collector grains. Song and Elimelech (1995) extended this analysis to particle deposition onto permeable surfaces, which allow the passage of fluid molecules through the pores but block the transport of colloidal particles, the deposition rate is controlled by the convection particle transport ($K_{dep} \propto U$). Hence, in this chapter below the critical velocity, the dependence of the measured deposition rate coefficient on water velocity ($U^{0.62}$, $U^{0.67}$) is between what is expected for deposition onto impermeable surfaces ($U^{1/3}$) and permeable porous surfaces ($U^{1.0}$).

The decreasing of deposition rate at highest flow velocities may be due to the existence of shear flow able to tear off particles and lower the deposition rate. At low flow rates the coarser suspended particles deposition is mainly controlled by the sedimentation where the suspended particles were subjected to the gravity and their velocity no longer is that of the fluid; thus, they can meet the filter medium. Other capture mechanisms such as capture in constriction and cavern sites occur too. Once the flow velocity was higher, the hydrodynamic force, especially drag force, is important and the relative significance of sedimentation with respect to bulk flow is low.

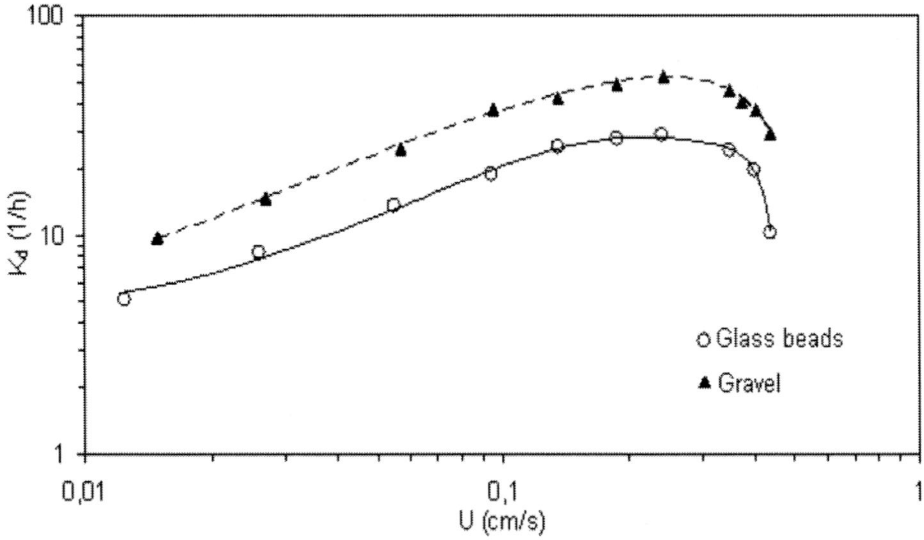

Figure 9. Suspended particles (SP) deposition kinetics as a function of water flow velocity (U).

The step-input experiments were performed in a short column (50 mm length and 4 mm diameter) using a suspension of kaolin particles (size ranging from 2 μm to 20 μm, with a mode of 7 μm). Figure 10 shows the breakthrough curve obtained in sand (100 μm – 500 μm) column experiment with a flow velocity of 0.045 cm/s. The breakthrough curves (figure 10) show two steps in the particle deposition. The particle deposition rate is initially high and almost no particles are detected in the effluent after the first pore volume. However, the reduced particle concentration in the effluent slowly increases and finally reaches a value close to 0.85 after about 10 pore volumes (figure 10), meaning that the particle deposition rate has dropped to very low value.

Particle deposition rates and attachment efficiencies can be determined from step-input or short-pulse column experiments. Following a step-input, particle breakthrough occurs at about one pore volume and the reduced breakthrough concentration C/C_0 reaches a plateau. For columns with a reasonably high peclet number ($Pe > 30$), the particle deposition rate coefficient can be obtained from:

$$k_d = -\frac{u}{L}\ln F \tag{11}$$

where L is the column length and F is the reduced particle concentration (C/C_0) after a stable plateau has been reached. The value of deposition rate coefficient deduced from the curve of Figure 10 is close to about 12 h^{-1}.

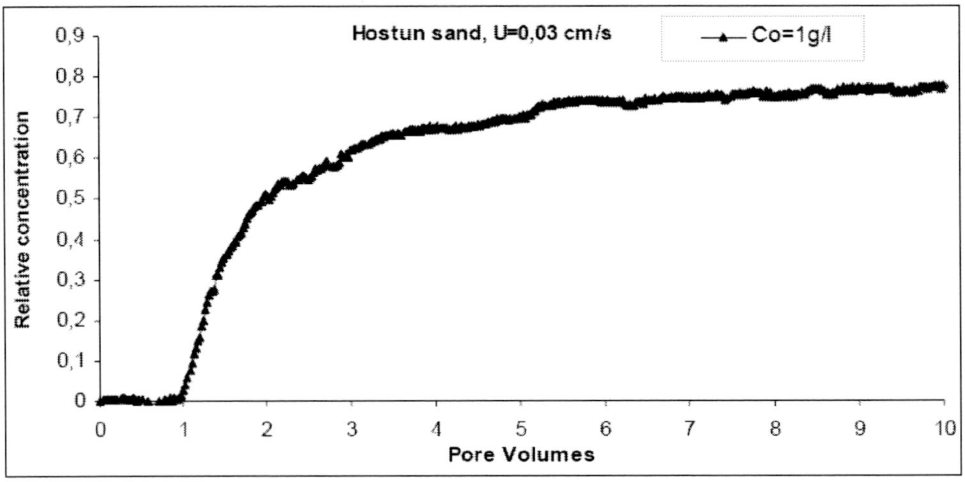

Figure 10. Breakthrough curve of step-input experiment in Hostun sand.

If one continue step-input injection of suspended particles, a steady state of deposition rate is reached (plateau) after a number of pore volumes greater than ten. The particle deposition rate is initially high and almost no particles are detected in the effluent after the first pore volume. However, the reduced particle concentration in the effluent slowly increases and finally reaches a value close to unity after about ten pore volumes (figure 11), meaning that the particle deposition rate has dropped to zero. When the influent is switched to water without particles, the particle concentration in the effluent drops quickly to zero without exhibiting significant tailing, a behaviour which can only be explained by irreversible deposition of the suspended particles. Thus, the matrix surfaces in the porous medium remain covered with deposited particles.

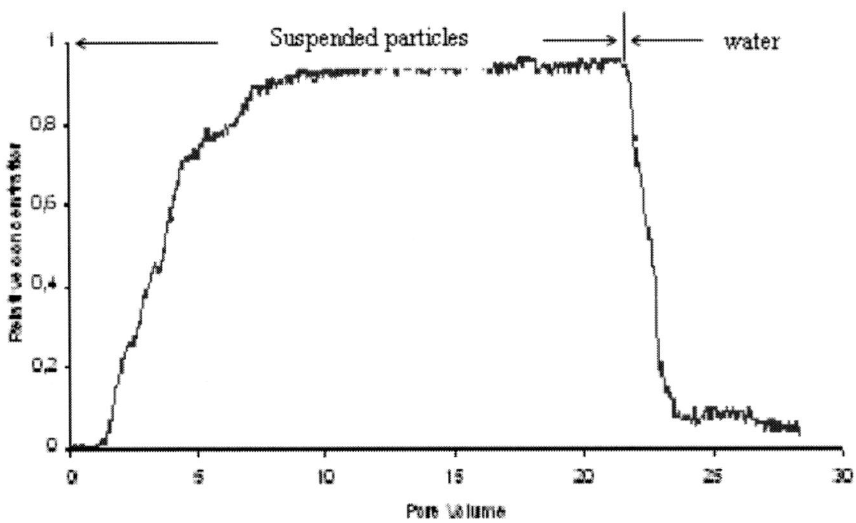

Figure 11. Transport of suspended particles in sand column exhibiting irreversible deposition.

7. Conclusions and Outlook

Laboratory research has provided evidence that suspended particles can be transported easily through subsurface porous media, faster than dissolved solute. Some issues of the transport and deposition of suspended particles (large size distribution) in saturated granular porous media were investigated in this chapter. The breakthrough curves of the pulse injections, at steady state flow, were completely described by the convection-dispersion equation with a first-order deposition rate. The analysis of breakthrough curves from step-input experiments provided an estimating of the deposition rate coefficient. The influence of particle flow velocity and pore structure was highlighted. This potential transport pathway should be considered in risk assessments of sites heavily contaminated. Transport behaviour comparison of suspended particles and conservative tracer, in two types of porous media, showed very remarkable changes. Suspended particles size exclusion effect is observed at high flow rates. The dispersion coefficient as found in the classical Fickian equation is a non linear function of flow velocity. This chapter concerned with transport and deposition rate, needs additional research on key processes such as the kinetics of particle release and deposition in natural heterogeneous porous media. The amounts of suspended particles recovered from the glass beads column are greater than those from the gravel column. At high velocity, the coarser suspended particles were most recovered. This difference in recovery rate is mainly due to the internal structure of the porous media: pore throats, retention sites, as well as the gravity and hydrodynamic forces. Considerable knowledge has accumulated over the years, highlighting the influence of ionic strength and surface chemistry on colloid transport. Such information can be used to assess the importance of physico-chemical effects on large particles transport. Little information is still available about the relative importance of polydisperse particles on particle-facilitated transport. Combined research at the field and laboratory scales is necessary will improve the ability to predict the fate of contaminants.

References

Ahfir, N-D.; Benamar, A.; Alem, A.; Wang, HQ. *Transp. Porous. Med.* 2008, DOI 10, 1007/s11242-008-9247-3.

Allaire, S.E.; Gupta, S.C.; Nieber, J.; Moncrief, J.F. *J. Contam. Hydrol.* 2002, 58, 299-321.

Auset, M.; Keller, A.A. *Water Resour. Res.* 2004, 40, W 03503.

Bear, J. *Dynamics of Fluids in Porous Media*. Elsevier: New York, 1972.

Benamar, A.; Wang, H.Q.; Ahfir, N.-D.; Alem, A.; Massei, N.; Dupont, J.P. *C. R. Geosci.* 2005, 337, 497–504.

Compère, F.; Porel, G.; Delay, F. *J. Contam. Hydrol.* 2001, 49, 1-21.

Cumbie D.H.; McKay L.D. *J. Contam. Hydrol.*1999, 37, 139-157.

de Marsily, G. *Quantitative Hydrogeology. Groundwater hydrology for engineers.* Academic Press INC: New York, US, 1986;

Elimelech, M.; Gregory, J.; Jia, X.; Williams, R.A. *Particle Deposition and Aggregation: Measurement, Modeling, and Simulation.* Butterworth-Heinemann: Oxford, UK, 1995.

Frey, J.M.; Schmitz, P.; Dufreche, I.; Gohr Pinheiro, I. *Transp. Porous Media.* 1999, 37, 25–54.

Gohr Pinheiro, I.; Schmitz, P.; Houi, D. *Chem. Eng. Sci.* 1999, 54, 3801–3813.

Grolimund, D.; Elimelich, M.; Borcovec, M.; Barmettler, K.; Kretzschmar, R.; Sticher, H. *Environ. Sci. Technol.* 1998, 32, 3562–3569.

Kretzschmar, R.; Barmettler, K.; Grolimund, D.; Yan, Y.D.; Borkovec, M.; Sticher, H. *Water Resour. Res.* 1997, 33, 1129–1137.

Massei, N.; Lacroix, M.; Wang, H.Q.; Dupont, J.P. *J. Contam. Hydrol.* 2002, 57, 21–39.

Ryan, J.N.; Elimelech,M. *Colloids Surf A: Physicochem. Eng. Asp.* 1996, 107, 1–56.

Sirivithayapakorn, S.; Keller, A.A. *Water Resour. Res.* 2003, 39, 1109.

Song, L.; Elimelech, M. *Colloids and Surfaces A.* 1993, 73, 49-63.

Song, L. ; Elimelech M. *J. Col. Interf. Sci.* 1995, 173, 165-180.

Wang, H.Q. *Transferts de matières en milieu saturé: outils mathématiques et modélisation numérique.* Université de Rouen : Rouen, FR, 2001 ;

Wang, H.Q.; Crampon, N.; Huberson, S.; *Garnier, J.M. J. Hydrol.*1987, 95, 143–154.

Wang, H.Q.; Lacroix, M.; Massei, N.; Dupont, J.P. *C. R. Acad. Sci. Paris Sci. Terre Planèt.* 2000, 331, 97–104.

In: Porous Media: Heat and Mass Transfer… ISBN: 978-1-60692-437-2
Editors: J. L. Acosta and A.F. Camacho

Chapter 9

MODELING THE RELATION BETWEEN COMFORT AND PROTECTION OF CBRN-SUITS

Paul Brasser*
TNO Defence, Security and Safety
PO Box 45, 2280 AA Rijswijk, The Netherlands

ABSTRACT

To optimize the balance between physiological burden and protection, models were developed to describe the important processes at hand. By using these models, the optimum requirements for CBRN-protective clothing can be established. These requirements will lead to optimum values of the various properties of the clothing. A large variety of models can be identified, which are important for such a tool. A selection of models is presented here: Agent vapor breakthrough through the clothing, Air flow around, through and underneath the clothing, Agent vapor concentration underneath the clothing, Agent deposition onto the skin.

The vapor breakthrough model, the pressure distribution model and the deposition model were verified experimentally.

INTRODUCTION

A soldier can experience the threat of chemical warfare agents, when working in the field. To protect himself against these vapors or liquids, he can wear NBC-clothing. The NBC-protective clothing, currently in use by military forces, is usually an air permeable carbon-based garment. The basis of the protection is adsorption of the toxic agent onto the carbon. The use of air-permeable materials in clothing reduces the thermal load offered by the clothing, by allowing air to flow through the material. At the same time the airflow can transport the toxic agents through the clothing, if the agent is not fully adsorbed by the carbon.

---

* Paul Brasser: +31 15 284 3303, paul.brasser@tno.nl

A way to identify the balance between physiological burden and protection is by modeling the important processes at hand and using these models to find the optimum. The system requirements following from this optimization will lead to requirements for subsystems and materials of the clothing. That way a simulation tool will be created, which can be used by several different types of users, for instance:

- Military planers for identifying the ideal suit for a specific mission
- Producers of CBRN clothing for creating the optimum clothing design
- Researchers for finding the ideal balance between physiological burden and protection
- Quality testing for identifying key parameters

To model the protective performance of air-permeable NBC-suits, several scales in the model have to be considered. Figure 1 illustrates the three scales:

- Micro scale: The protective performance of the material itself.
- Meso Scale: The effect of air flow around body parts onto the deposition on the skin
- Macro scale: The model of the whole system (suit)

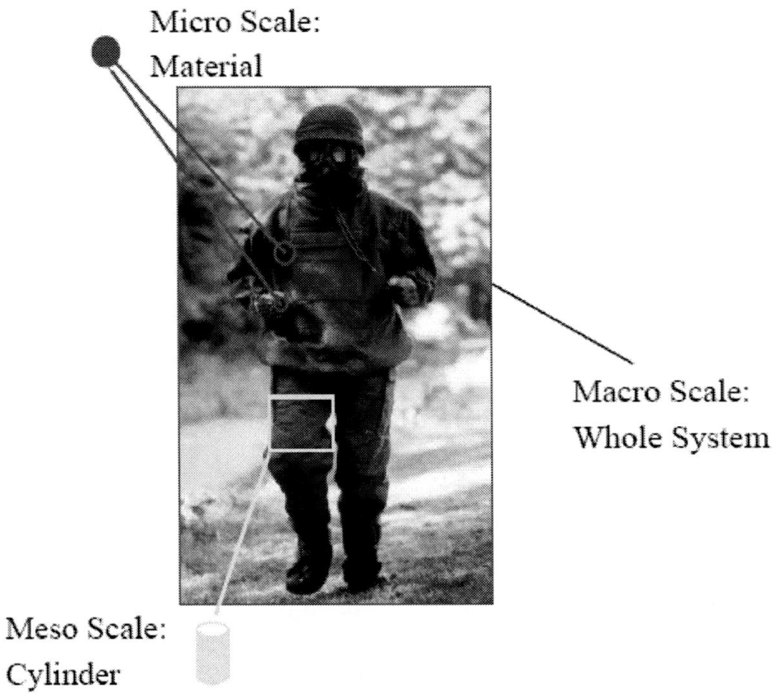

Figure 1. The process scales, important for the protection of a NBC-suit.

Several processes occur when wearing protective clothing. To investigate both the protective performance and the comfort parameters, the following processes can be identified:

- Air flow around, through and underneath the clothing
- Agent vapor breakthrough through the clothing
- Agent vapor concentration underneath the clothing
- Agent deposition onto the skin
- Water vapor penetration through the clothing
- Water vapor concentration underneath the clothing
- Evaporation and condensation of water from/in the clothing and from/ onto the skin
- Temperature distribution through the clothing, underneath the clothing
- Skin temperature and core temperature

To model all these processes underneath clothing of a human in full detail, use computational fluid dynamics (CFD) is required. For instance the modeling of the air flow distribution around cylinders was described by using CFD [1-4]. Parameters like the air velocity distribution, the concentration distribution and the temperature distribution were analyzed previously. Several types of turbulence models were examined. In figure 2 an example of results of the very time-consuming DNS calculations are shown.

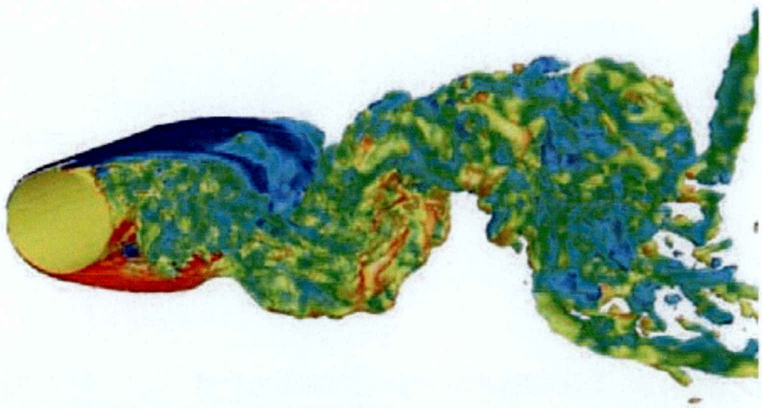

Figure 2. CFD calculations (DNS) of the airflow around a cylinder.

For various reasons, it is convenient to have a model, which can predict the behavior of these processes without using CFD, without high loss of accuracy (for instance because of integration with other models, easiness of use, easiness to change parameters, calculation time (when for instance DNS is used).

If homogeneous perpendicular flow of the outside wind is assumed and the air flow underneath the clothing is modeled one-dimensionally, the processes can be solved analytically. Therefore all these processes have to be modeled by basic one-dimensional physical transport equations.

PROCESSES

In the introduction a list of important processes was identified. In this chapter, the following processes will be discussed in more detail:

- Agent vapor breakthrough through the clothing
- Air flow around, through and underneath the clothing
- Agent vapor concentration underneath the clothing
- Agent deposition onto the skin

More extended descriptions of these processes and their subsequent models can also be found elsewhere [5-7].

Agent Vapor Breakthrough through Clothing

If the flow distribution around, through and underneath the clothing is known, a concentration distribution can be derived. The first step will be to calculate the agent vapor breakthrough through the protective clothing.

In theoretical analysis only NBC-protective clothing materials of the carbon bead type have been taken into account. In this type of protective clothing the chemical filtration is based on a single layer of small activated carbon spheres adhered a carrier fabric (figure 3).

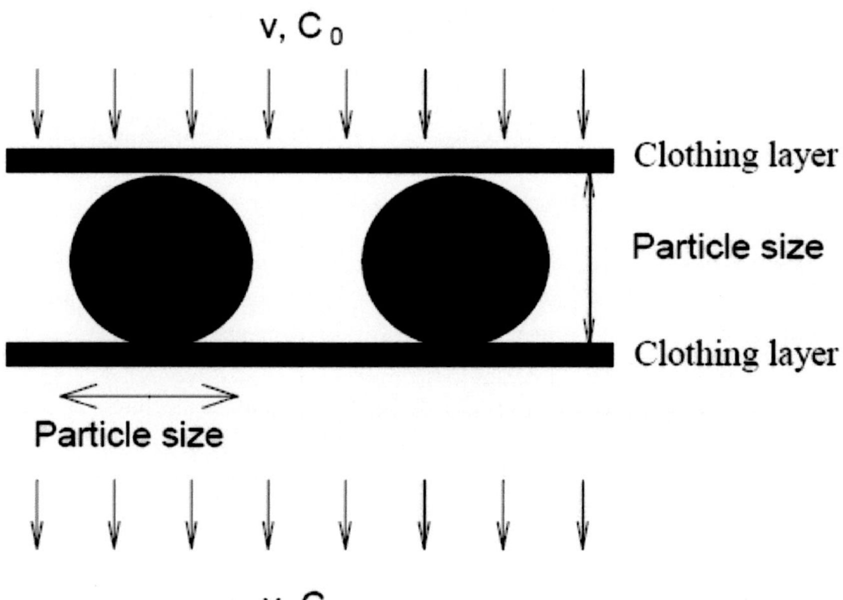

Figure 3. A schematic representation of a layer of NBC-clothing material.

When an activated carbon filter is challenged by a chemical agent vapor flow, the breakthrough curve of the effluent concentration against time is typically S-shaped. Typical for carbon bead type fabrics is an initial step in the breakthrough curve. Immediately after exposure a very small breakthrough concentration occurs that is roughly constant over a certain period of time (typically a few hours for NBC-clothing material).

Vapor Breakthrough Model

Several authors have studied the breakthrough of vapor through carbon filters [8,9]. A commonly used equation, which describes the vapor concentration, C, inside the filter as a function of the axial position, z, in the filter, is:

$$\frac{\partial C}{\partial t} = D\frac{\partial^2 C}{\partial z^2} - v_{mat}\frac{\partial C}{\partial z} - k_g\frac{(1-\varepsilon)}{\varepsilon}\frac{3}{r}(C - C_s) \tag{1}$$

Refer to the symbols section for the definition of the various symbols. Or by introducing the dimensionless parameters:

$$\xi = \frac{(C - C_{ini})}{(C_0 - C_{ini})} \qquad \tau = \frac{t}{t_{50}} \qquad \chi = \frac{z}{2r}$$

$$Sh = \frac{rk_g}{D} \quad \mathrm{Re} = \frac{\rho v_{mat} 2r}{\eta} \quad Sc = \frac{\eta}{\rho D} \quad Fo = \frac{Dt_{50}}{4r^2} \tag{2}$$

Equation 1 can be transformed to:

$$\frac{1}{Fo}\frac{\partial \xi}{\partial \tau} = \frac{\partial^2 \xi}{\partial \chi^2} - \mathrm{Re}\, Sc\frac{\partial \xi}{\partial \chi} - 12\frac{(1-\varepsilon)}{\varepsilon}Sh(\xi - \xi_s) \tag{3}$$

Although this model is not completely applicable for NBC-protective clothing, an estimate is needed for the initial breakthrough concentration. In the initial stages of the breakthrough, the vapor concentration in the carbon, C_S, is almost 0, because no adsorption has taken place yet. Experimentally it was found that the initial breakthrough concentration through NBC-protective clothing material remains constant for several hours before any significant change in the breakthrough concentration occurs. Therefore, during the initial moments, another assumption can be made: the initial breakthrough concentration remains constant. Mathematically this means that:

$$\frac{\partial \xi}{\partial \tau} = 0 \tag{4}$$

Thus:

$$\frac{\partial^2 \xi}{\partial \chi^2} - \mathrm{Re}\, Sc\frac{\partial \xi}{\partial \chi} - 12\frac{(1-\varepsilon)}{\varepsilon}Sh(\xi - \xi_s) = 0 \tag{5}$$

The two boundary conditions of this differential equation are:

- The concentration, C, is equal to the challenge concentration, C0, at the inlet of the filter (z = 0). Thus, $\xi = 1$ at $\chi = 0$.
- The concentration at the outlet, Cini, must be lower than the inlet concentration, C0. Thus, $\xi < 1$ at $\chi = 1$.

Sh is a function of the air velocity. Wakao et al. [10] propose an empirical relation between Sh and Re, based on experimental data, which was published by various authors in the past. This equation is valid in the range 3 < Re < 10000. In the case, which is described here, Re is always smaller than 3 because both the carbon particles are very small (0.5 mm diameter) and the air velocity is quite small (maximum 0.08 m/s). For this highly viscous flow regime, Sh becomes equal to 2 [11], which is assumed here.

The particle layer porosity of the carbon particles follows from the carbon load, Lo, and the density of the carbon particles, ρ_c, since one minus the porosity is occupied by the carbon:

$$(1-\varepsilon)=\frac{Lo}{2r\rho_c} \tag{6}$$

Using the boundary conditions, differential equation 3 can be solved for the initial stage of the breakthrough process. This gives the initial breakthrough concentration:

$$\xi_s=(\xi_s-1)\exp\left(\frac{\mathrm{Re}\,Sc}{2}\left(1-\sqrt{1+\frac{96\,\mathrm{Re}^{-2}\,Sc^{-2}}{\left(\frac{2r\rho_c}{Lo}-1\right)}}\right)\right) \tag{7}$$

or in dimensional form:

$$C_{ini}=C_0\exp\left(\frac{vr}{D}\left(1-\sqrt{1+\frac{24}{\left(\frac{2r\rho_c}{Lo}-1\right)}\left(\frac{D}{rv}\right)^2}\right)\right) \tag{8}$$

Another characteristic parameter of a breakthrough curve is the time at which 50% breakthrough occurs, which is determined by the adsorption capacity of the carbon. A breakthrough curve is typically S-shaped (mimicking an error function in time). It is assumed that the shape of the dimensionless breakthrough curve for times smaller than the 50% breakthrough is the inverse of that above the 50% breakthrough (which is the case for a perfect S-shaped curve). In that case, the amount of vapor adsorbed on the carbon is equal to the dose at which the carbon material has been exposed, $(C_0 - C_{ini})t_{50}$, times the velocity of the vapor through the material, v_{mat}. Taking into account the carbon load, the 50% breakthrough time can be predicted by:

$$t_{50} = \frac{Lo\,q}{(C_0 - C_{ini})v\rho_c} \tag{9}$$

Equilibrium between the gas and solid phase concentrations is described with the Dubinin-Radushkevich adsorption isotherm [8,9]:

$$q = q_{\max} \exp\left(-\left(\frac{R_{gas}T}{\beta E_0} \right)^2 \ln^2\left(\frac{C}{C_{sat}} \right) \right) \tag{10}$$

Thus, for every challenge concentration and air velocity, the 50% breakthrough time can be calculated.

For carbon filters equation (3) can be used to calculate the breakthrough concentration as a function of time, in combination with an equation for the adsorbed vapor inside the carbon. NBC-protective clothing material consists of only one layer of carbon particles. Equation (3) does not describe the process in a one-particle layer situation appropriately. Differences, compared to a packed bed, that have to be taken into account, are: the particle layer porosity changes over the thickness and a carbon particle experiences another vapor concentration at the inlet than at the outlet. By CFD these problems can probably be solved. Another approach is to find an analytical approximation, which describes the breakthrough concentration as a function of time. Though it is just argued that equation (3) does not apply completely in this case, it is still able to describe the experimental behavior quite well for the initial stage of the breakthrough curve (equation (8)). To describe the breakthrough concentration at later stages of the breakthrough curve, a time dependency is needed (equation (3) does not give proper predictions for these later stages). A function, which fits the experimental data, is a closed form expression and resembles the analytical solution of equation (3), can be used for this purpose. Solving partial differential equations analytically is only possible in certain cases, none of which apply here. However, it is known, that the analytical solution typically has the shape of an error function as a function of time and furthermore the initial breakthrough concentration and the 50 % breakthrough time are also known ($\xi = 0$ at $\tau = 0$ and $\xi = 0.5$ at $\tau = 1$). This will lead to the equation:

$$\xi = \frac{1}{2} + \frac{1}{2} erf\left(\frac{(\tau - 1)}{\sigma\sqrt{2}} \right) \tag{11}$$

when the initial breakthrough concentration, the 50 % breakthrough time and the steepness of the error function, σ, are known, the total breakthrough curve can be plotted. The value for the steepness can follow from fitting the (dimensionless) experimental breakthrough data. That way the total breakthrough curve is described semi-empirically.

Vapor Breakthrough Measurements

The protective performance of bead type NBC-protective clothing material was determined using a mustard gas vapor challenge test [12]. A forced flow of a nitrogen stream (relative humidity<10%, temperature 25◦C) with mustard gas vapor was created through the material samples. The concentration of the penetrated agent as a function of time was measured using a gas chromatograph (Chrompack_R CP9001; with 10-meter column CP-Wax-52 CB) with a flame ionization detector. Nitrogen was used as purge flow (35 ml/min) and the column temperature was set at 170◦C.

The influence of the air velocity through the material on the penetration of mustard vapor was determined by conducting breakthrough tests at different air velocities 2, 4, 6, and 8 cm/sec while the challenge concentration was kept constant at approximately 11 mg/m3. Another set of breakthrough experiments was performed at a constant air velocity of 5.7 cm/sec while varying the challenge concentration between 4.45, 10.00, 17.63, and 43.88 mg/m3.

Important parameters in this breakthrough measurements are:

- the initial breakthough concentration
- the 50 % breakthrough time
- the total breakthrough concentration curve

These parameters will be discussed in more detail below.

Vapor Breakthrough Results

Initial Breakthrough

The calculated and measured initial breakthrough concentration is plotted in figure 4 as a function of the air velocity through the material.

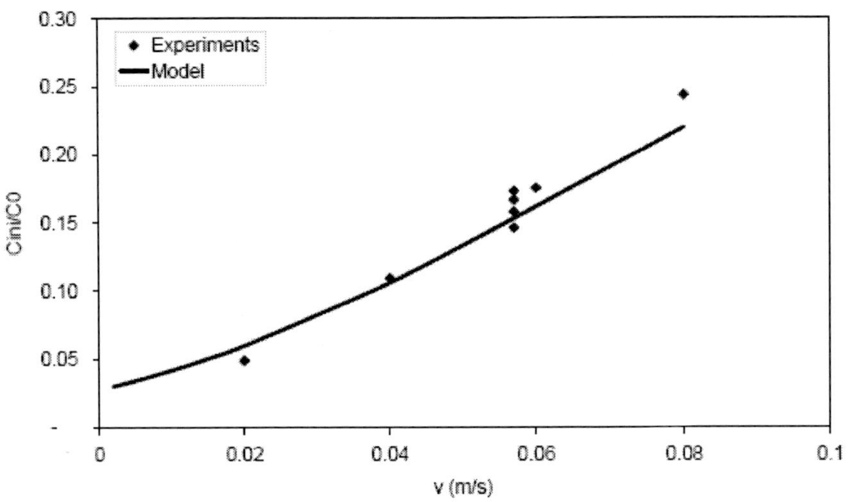

Figure 4. The initial breakthrough concentration as a function of the air velocity through the textile.

Figure 4 shows a good agreement between the calculated and the measured initial breakthrough concentration. Furthermore an increase as a function of the air velocity is found.

Experimentally an almost linear dependence between the air velocity and the relative initial breakthrough concentration was found. Higher air velocity results in less contact time between the air and the carbon, and therefore a higher initial breakthrough concentration occurs.

By using equation (10), the initial breakthrough concentration was calculated in the same velocity range. The model shows a slight exponential dependence, which can, in the specific air velocity range, easily be approximated by a linear dependence between the relative initial breakthrough concentration and the air velocity.

50% Breakthrough

The 50 % breakthrough time was calculated with the model and measured experimentally (figure 5). As can be seen, the predicted 50 % breakthrough time shows good agreement with the measured one.

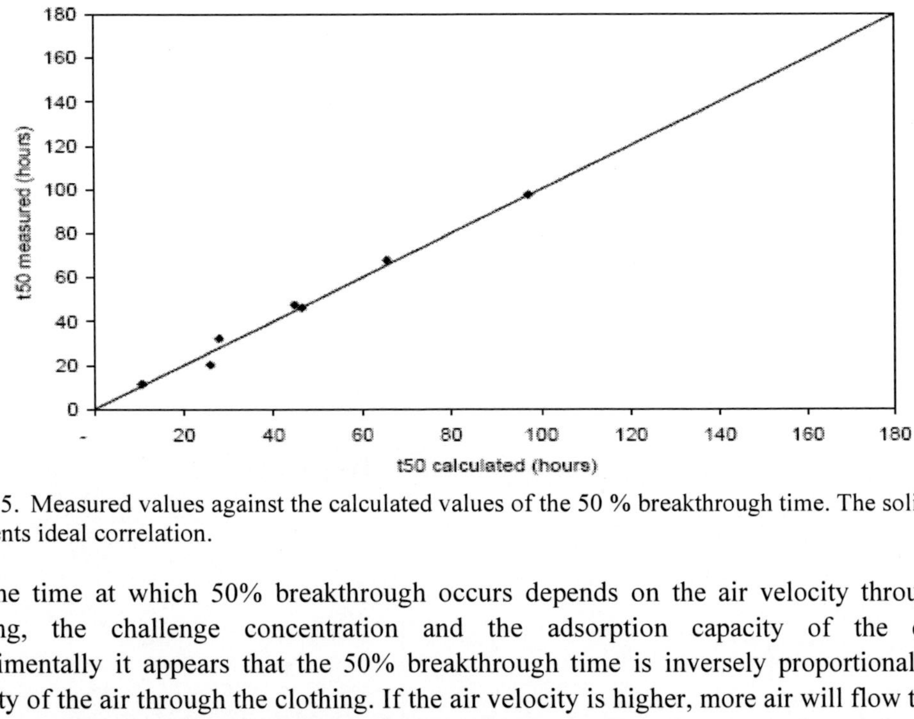

Figure 5. Measured values against the calculated values of the 50 % breakthrough time. The solid line represents ideal correlation.

The time at which 50% breakthrough occurs depends on the air velocity through the clothing, the challenge concentration and the adsorption capacity of the carbon. Experimentally it appears that the 50% breakthrough time is inversely proportional to the velocity of the air through the clothing. If the air velocity is higher, more air will flow through the clothing during a certain time period. Thus more vapor must be adsorbed onto the carbon. This means that the carbon will reach its maximum adsorption capacity earlier, resulting in a decrease of the 50% breakthrough time.

Changing the challenge concentration will effect the 50 % breakthrough time in two ways. If the challenge concentration increases, the 50 % breakthrough time will decrease, since the carbon has to adsorb more vapor in the same time interval. On the other hand, the total adsorption capacity will increase according to its adsorption isotherm, which will result in an increase in the 50 % breakthrough time. The first effect is larger than the second effect;

thus the overall effect will be a decrease in the 50 % breakthrough time if the challenge concentration increases.

Total Breakthrough

The total breakthrough curve of sulphur mustard through CBRN-clothing material was measured experimentally as a function of time, the challenge concentration and the air velocity through the material (figure 6) by Brasser [5].

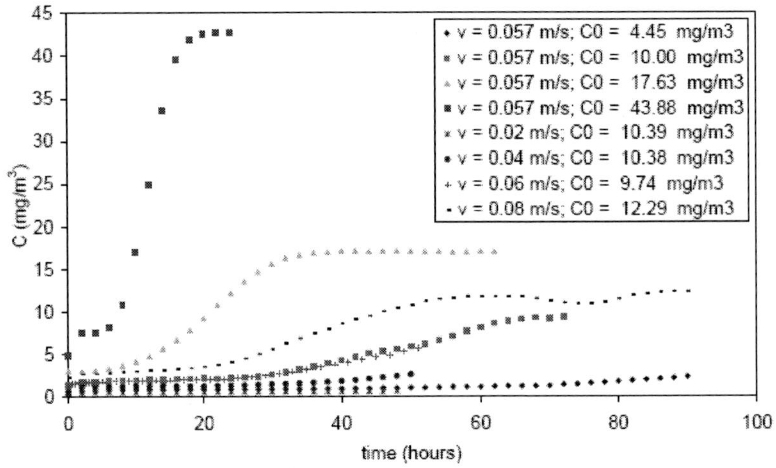

Figure 6. The vapor breakthrough concentration through NBC-protective clothing material as a function of time.

As figure 6 shows, both the initial breakthrough concentration and the 50 % breakthrough time strongly depend on the challenge concentration and the air velocity through the material.

Both the calculated initial breakthrough concentration and the calculated 50 % breakthrough time were used make the measured breakthrough curve dimensionless. When the dimensionless concentration parameter ξ is plotted as a function of dimensionless time, τ, a remarkable collapse of the experimental data is found (figure 7).

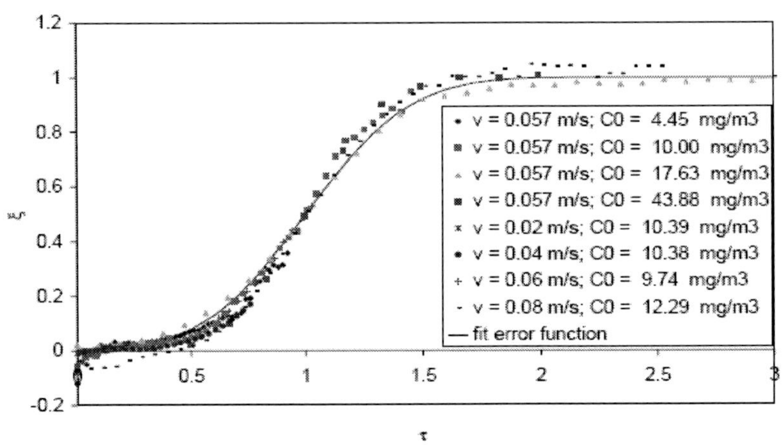

Figure 7. The dimensionless vapor breakthrough through NBC-protective clothing material as a function of dimensionless time.

Thus, the steepness of the breakthrough curve is independent of parameters like vapor concentration, air velocity, and time. All the breakthrough curves that were used to fit the steepness of the error function, σ, of equation (11) (regression coefficient $r^2 = 0.995$):

$$\xi = \frac{1}{2}\left(1 + erf\left(2(\tau - 1)\right)\right) \tag{12}$$

or:

$$C = C_{ini} + \left(C_0 - C_{ini}\right)\frac{1}{2}\left(1 + erf\left(2\left(\frac{t}{t_{50}} - 1\right)\right)\right) \tag{13}$$

This equation describes the breakthrough concentration of sulphur mustard through bead type NBC-protective clothing material.

Air Flow Around, through and Underneath Clothing

To model the air flow distribution around a human limb, a simplification has been made by assuming this to be a two dimensional problem. The body part is represented by a cylinder which is placed in airflow. At the front of the (dressed) cylinder the wind will penetrate the clothing and at the backside it will flow back to the environment. If only homogeneous perpendicular flow of the outside wind is assumed and the air flow underneath the clothing is modeled only one-dimensionally, the process can be solved without using CFD. A schematic picture of the set-up is shown in figure 8.

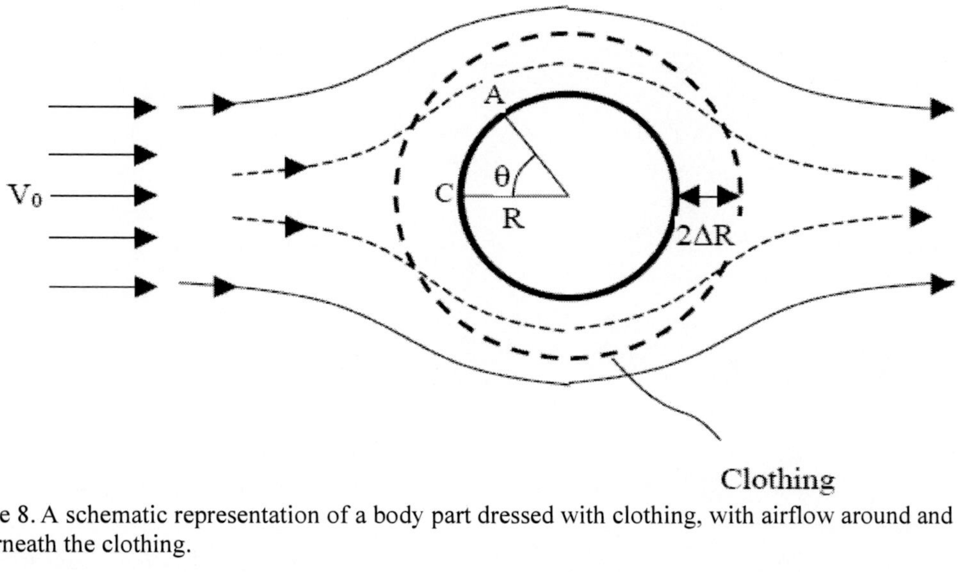

Figure 8. A schematic representation of a body part dressed with clothing, with airflow around and underneath the clothing.

The wind is blowing against the clothing around the cylinder with a velocity v_0. At the front of the cylinder, at point C (the stagnation point) the wind will partly flow through the porous clothing and will continue to flow underneath the clothing. At different points around the cylinder (different angles from the stagnation point, for example point A), the velocity through the clothing will be different, due to the fact that the pressure around the cylinder varies. Up to a certain angle, the wind will flow into the air gap, but at larger angles, the wind will flow out again. This process depends on the pressure in the air gap underneath the clothing, which is also a function of the position around the cylinder.

Air flow and Pressure Distribution Model

The flow distribution between the clothing material and the cylinder can be approximated by a fully developed flow distribution between two infinitely large plates. In figure 9 two plates are shown with a flow of a fluid between them.

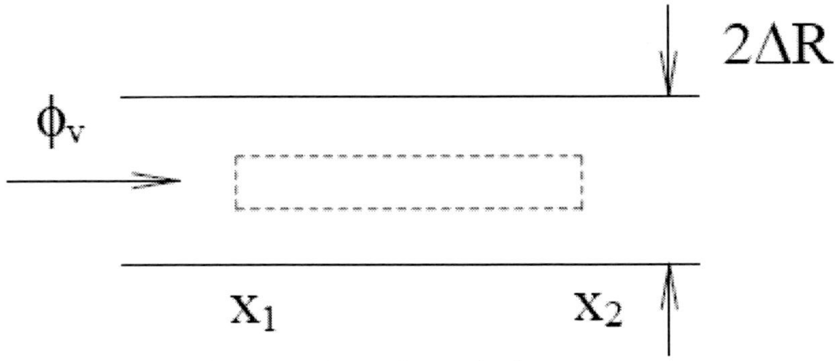

Figure 9. Airflow between two plates. A volume element is shown.

The distance between the two plates equals 2ΔR. The length of the plates (in the depth of the figure) equals L. Poisseulle flow is assumed, thus the velocity of the air can be described by:

$$v_r = \frac{1}{2\eta}\left(-\frac{\partial P}{\partial x}\right)\left((\Delta R)^2 - r_g^2\right) \tag{14}$$

Note that the symbols are defined in the Notation section. The flow between the plates can be calculated by:

$$\phi_v = 2\int_0^{\Delta R} L v_r dr = \frac{2}{3}\frac{L}{\eta}\left(-\frac{dP}{dx}\right)(\Delta R)^3 \tag{15}$$

This equation will give the flow between two parallel plates. This relation can be modified for the flow of air between clothing and a cylinder by assuming that the radius of the cylinder is large in comparison to the size of the air gap. In that case the next approximation can be used:

$$dx \approx Rd\theta \tag{16}$$

The flow between the clothing and the skin can thus be calculated by:

$$\phi_v(\theta) = \frac{2}{3}\frac{L}{\eta}\left(-\left.\frac{dP_{in}}{d\theta}\right|_\theta\right)\frac{(\Delta R)^3}{R} \tag{17}$$

The average velocity of the air between the clothing material and the cylinder is:

$$v_{air}(\theta) = \frac{\phi_v(\theta)}{2\Delta RL} = \frac{1}{3\eta}\left(-\left.\frac{dP_i}{d\theta}\right|_\theta\right)\frac{(\Delta R)^2}{R} \tag{18}$$

The velocity of the air through the clothing depends on the pressure difference over the clothing. For high velocities (for instance through parachutes) the relation is quadratic [13,14]. For lower velocities (which is the case here) this relation becomes linear. Therefore a Darcy type of flow is assumed [15]. The law of Darcy describes the velocity through a filter as:

$$v_{mat} = K_D \frac{l}{\eta}\Delta P \tag{19}$$

In the case of air, K_D, l and η can be combined into the air permeability, Γ. Thus the velocity through the clothing is calculated by:

$$v_{mat} = \Gamma\Delta P \tag{20}$$

Due to this air velocity, the flow underneath the clothing will change [16,17]:

$$d\phi_{v,2D} = 2\Delta R v_{mat} dx = 2\Delta R v_{mat} R d\theta \tag{21}$$

By inserting equation (17) and (19) into equation (21), the next partial differential equation can be deduced:

$$\frac{2}{3\eta}\left(-\left.\frac{d^2P_{in}}{d\theta^2}\right|_\theta\right)\frac{(\Delta R)^3}{R} = \Gamma R\Delta P = \Gamma R(P_{out} - P_{in}) \tag{22}$$

If the pressure outside of the clothing is known, the pressure underneath the clothing can be calculated by this equation. The velocity of the air through and underneath the clothing will follow from equation (18) and (19). The pressure outside the clothing can be calculated if the airflow is very laminar and no separation point of the boundary layer is present [18]. However, if the airflow is turbulent (which is usually the case), the outside pressure can not be calculated as a function of the angle. It is though possible to use experimental (outside) pressure values to calculate the inside pressure. This approach is used here.

Pressure Distribution Measurements

To measure the pressure distribution, a cylinder as a model for a body part was placed in a test chamber. In this test chamber airflow can be generated with a defined wind velocity. A low and a high wind velocity were studied (1.4 m/s and 7.2 m/s). The wind velocity is measured continuously during the experiment with a hot-wire anemometer, which is mounted at a distance of circa 20 cm at the side of the cylinder (figure 10).

Figure 10. NBC-fabric around a cylinder (model for a body-part).

The flow causes pressure differences around the cylinder, which are detected by micro tubes, mounted inside the cylinder. The pressure tubes are separated from each other by an angle of 10° around the cylinder [19,20] and they are connected to a pressure-difference scanning device (NetScanner Model 9016), which measures the pressure difference between the sample tube and a reference tube. The reference tube is placed in the same room (thus has the same temperature). Effects of airflow onto the reference pressure are minimized by placing the sampler perpendicular to the air flow and by covering it with a thin (flexible) layer of tape. The pressure results are processed in a data application module (NetScanner Model 90DB), which in turn is connected to a PC to analyze the data.

Every measurement consists of the average value of the measured pressure during 1 min, which reduces the variations due to turbulent flow. To account for systematical errors, every measured pressure was subtracted by the measured value at a wind speed of 0 m/s at the same location.

The cylinder was covered by a layer of clothing, which is draped at a defined distance from the cylinder wall. The distance wall - clothing is set by rings around the cylinder, which in turn are used to mount the clothing on. The thickness of the rings determines the distance cylinder wall – clothing (the air gap). The air gap size was kept constant at 4 mm during all experiments. Though the air can cause the clothing to be deformed, as described in the variable air gap size chapter, it is presumed that this does not happen.

The pressure distribution around the cylinder was measured for both a bare cylinder and a cylinder which is dressed with clothing. For the theoretical pressure calculations in the air gap it is assumed that the pressure distribution around a bare cylinder resembles the pressure distribution at the outside of the clothing of a dressed cylinder.

Two types of clothing were used, one with a low and another one with a high air permeability ($7.3*10^{-3}$ and $6.2*10^{-4}$ m/(Pa s)).

Pressure Distribution Results

Outside Pressure Distribution

Pressure distributions around a bare cylinder were measured as descried. The measured distributions were normalized by dividing them by the measured pressure at the stagnation point. In figure 11 these measured distributions are given for the two wind speeds.

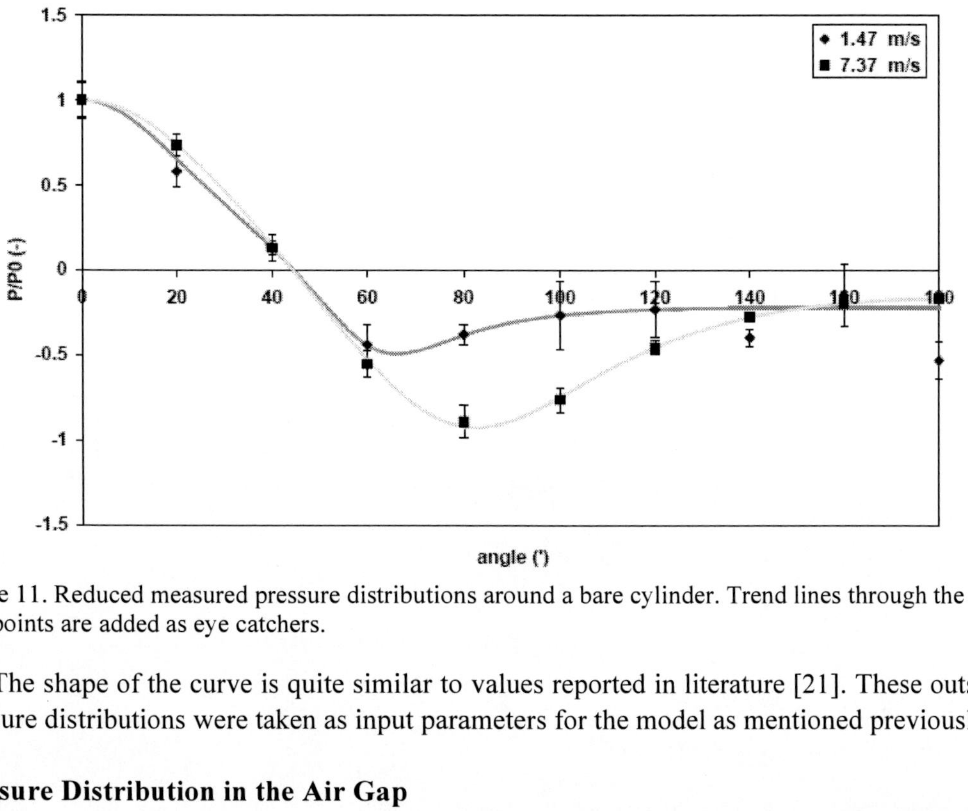

Figure 11. Reduced measured pressure distributions around a bare cylinder. Trend lines through the data points are added as eye catchers.

The shape of the curve is quite similar to values reported in literature [21]. These outside pressure distributions were taken as input parameters for the model as mentioned previously.

Pressure Distribution in the Air Gap

Brasser also measured the pressure distribution in the air gap [6]. The calculated pressure distributions in the air gap are shown together with the measured values in figure 12 and figure 13.

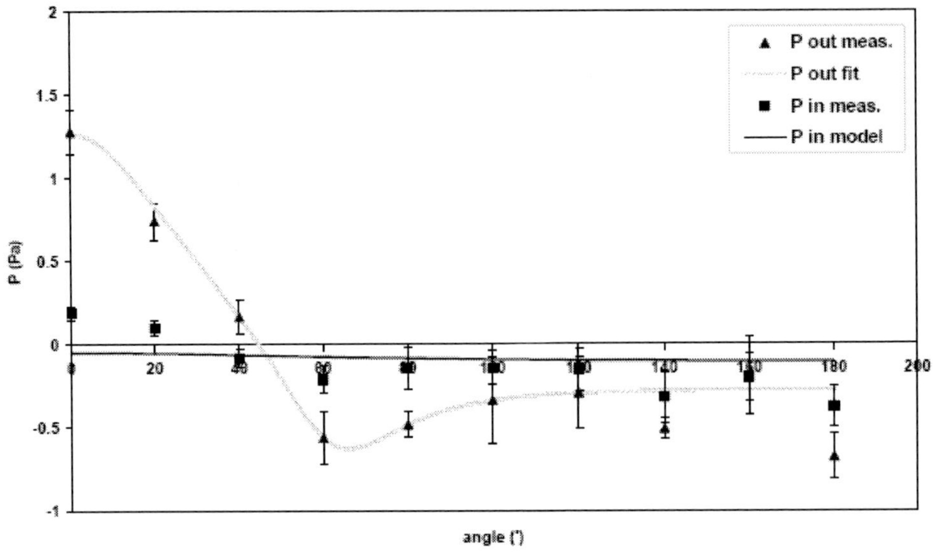

Figure 12. Pressure distribution around a bare cylinder and under clothing with high air resistance at a low wind velocity.

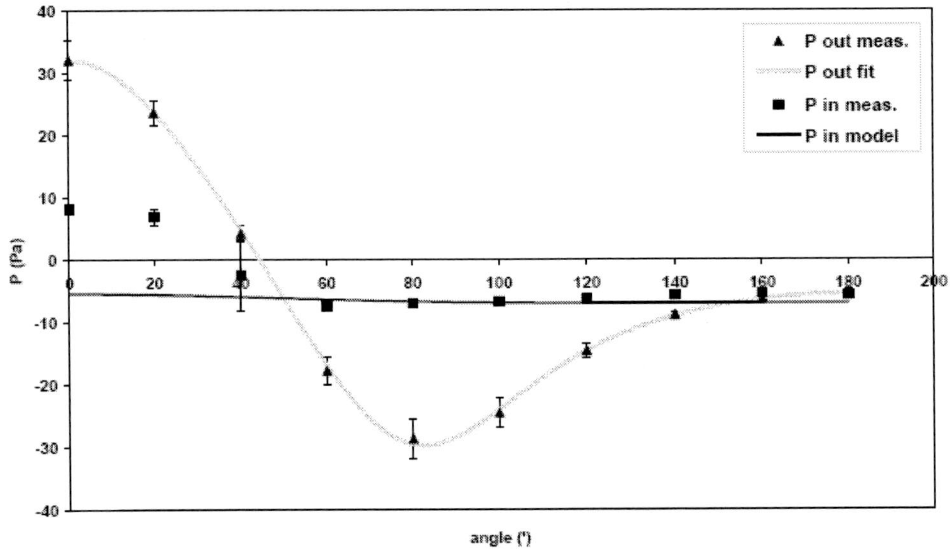

Figure 13. Pressure distribution around a bare cylinder and under clothing with low air resistance at a high wind velocity.

The air gap pressure (underneath the clothing) does not vary much for the two types of clothing (high and low air permeability). A change in the velocity has much more effect on both the inside and the outside pressure. The calculated pressure underneath the clothing shows good agreement with the measured values. In the case of highly air-permeable clothing at high air velocity, in the front of the cylinder the measured pressures are somewhat higher than the calculated values. A possible explanation for this observation can be that the size of the air gap at the front of the cylinder is not exactly the same as the value which is used in the calculations, because the pressure of the wind will decrease this size somewhat (see variable air gap size results).

Air Velocity through the Clothing

The air velocity through the clothing can be calculated from the measured outside pressure and pressure in the air gap. The velocity distributions, predicted by the model, are also shown (figure 14).

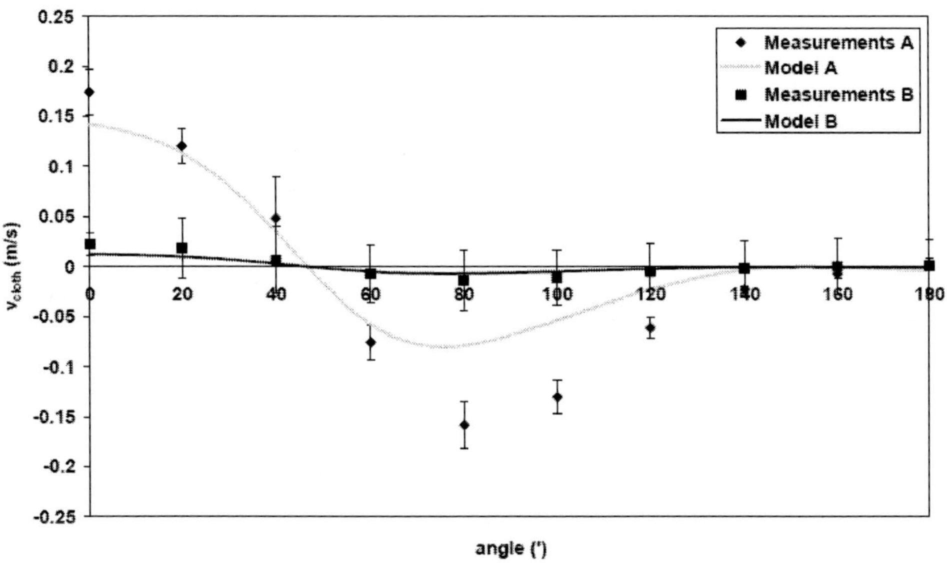

Figure 14. The velocity distribution through NBC-clothing as a function of the angle around the cylinder.

The model predicts a somewhat higher value for the air velocity through the clothing. This is presumably the result of the effect, which is discussed previously, regarding a variable air gap size.

The velocity through the clothing scales linearly with the air permeability. Thus air permeability has a large effect onto the velocity through the clothing. The air velocity through the clothing changes largely with the wind velocity.

Air Velocity in the Air Gap

The air velocity in the air gap between the cylinder and the clothing is related to the pressure difference in this air gap over the angle (equation (17)). Since the variation in the pressure measurements is larger than the difference between the values of two angles, it is not realistic to calculate the air velocity underneath the clothing from these measurements. Theoretically expected values can however be calculated (figure 15).

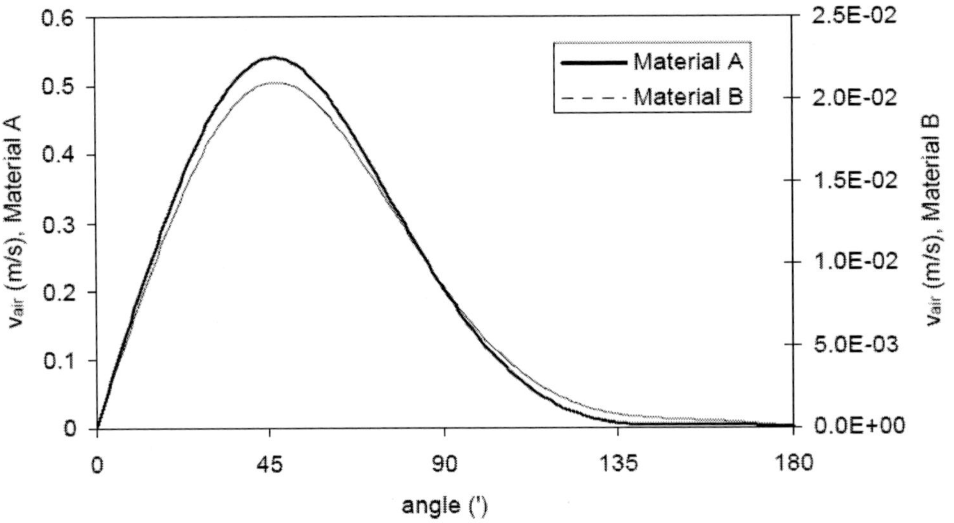

Figure 15. The air velocity underneath the clothing around a body part as a function of the angle.

The air velocity underneath the clothing is a lot higher than the air velocity through the clothing (compare for instance figure 14 and figure 15). At a high wind flow combined with high air permeability, the air velocity almost equals the wind velocity at a certain angle.

Vapor Concentration and Deposition Distribution Underneath Clothing

The breakthrough concentration model and the airflow distribution model can be used as input in the next step to model the concentration distribution underneath the clothing and the local mass deposition on the cylindrical surface. For this step it is assumed that:

- all the mass of the chemical agent which reaches the cylinder surface is adsorbed.
- the mass transport to the surface is described by the penetration theory [11,22].

The first assumption implies that the concentration on the surface of the cylinder is always zero. A continuity equation (a mass balance for a slice of air underneath the clothing) describes the vapor concentration underneath the clothing as a function of the angle θ with respect to the wind direction around the cylinder (figure 16).

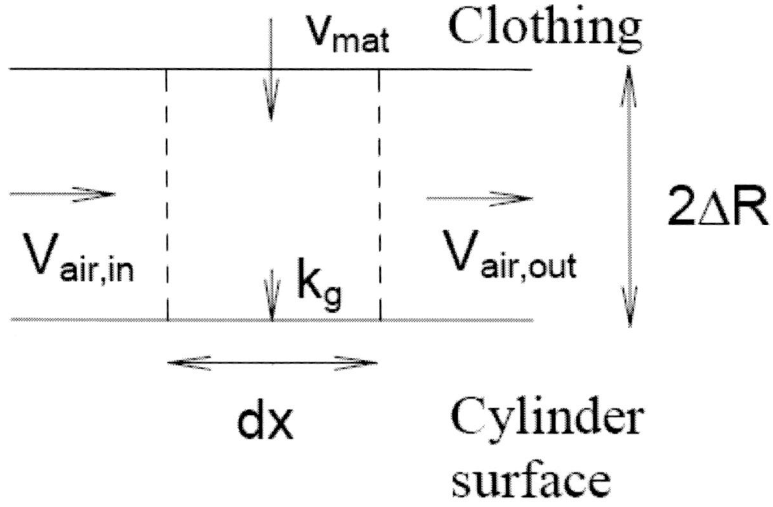

Figure 16. A volume element of air between the clothing and the surface of the cylinder.

Vapor Concentration Model

The continuity equation will contain both convection and diffusion terms and is described by a partial differential equation, which describes the concentration between the clothing and the skin:

$$\frac{dC}{dt} = D\frac{d^2C}{dx^2} - v_{air}\frac{dC}{dx} + \frac{v_{mat}}{2\Delta R}(C_0 - C) + \frac{\varepsilon D}{2\Delta R d_{mat}}(C_0 - C) - \frac{k_g}{2\Delta R}C \quad (23)$$

Equation 1 describes the concentration between two parallel plates. This relation can be modified for the flow of air between clothing and a cylinder by assuming that the radius of the cylinder is large in comparison to the size of the air gap. In that case an approximation can be used for dx:

$$dx \approx Rd\theta \quad (24)$$

Equation 23 then changes into equation 25.

$$\frac{dC}{dt} = \frac{D}{R^2}\frac{d^2C}{d\theta^2} - \frac{v_a}{R}\frac{dC}{d\theta} + \frac{v_{mat}}{2\Delta R}(C_0 - C) + \frac{\varepsilon D}{2\Delta R d_{mat}}(C_0 - C) - \frac{k_g}{2\Delta R}C \quad (25)$$

The first term at the right represents the diffusion underneath the clothing, the second term the convection underneath the clothing, the third the convection through the clothing, the fourth the diffusion through the clothing and the fifth the deposition onto the surface of the cylinder. The boundary conditions of this equation are given in equation 26.

$$\left.\frac{dC}{d\theta}\right|_{\theta=0} = 0$$

$$\left.\frac{dC}{d\theta}\right|_{\theta=\pi} = 0 \tag{26}$$

$$\left.C\right|_{t=0} = 0$$

$$\left.C\right|_{t=\infty} = C_d$$

Vapor Deposition onto the Skin

Agent, which is present as vapor in the air gap between the clothing and the cylinder surface, can deposit onto the surface of the cylinder. The amount of deposition onto this surface will determine the effects of this agent. For simplicity, the penetration theory is assumed, which implies that a boundary layer will be created near the surface in the air (though not totally consistent for Poiseulle flow). The concentration distribution in this boundary layer is supposed to be linear. Usually the thickness of the boundary layer, δ, is unknown. It is customary to combine δ with the diffusion coefficient D to calculate kg, equation 27 [11]:

$$k_g = \frac{D}{\delta} \tag{27}$$

The mass transfer coefficient, k_g, is unknown too. The thickness δ is a function of the air velocity underneath the clothing, the covered distance along the cylinder surface, x, and the thickness of the air layer between the clothing and the cylinder surface, 2ΔR. Thus the thickness of the air gap determines the maximum boundary thickness. It is customary to write these dependencies in the dimensionless numbers Sh (Sherwood), Re (Reynolds) en Sc (Schmidt). The penetration theory gives equation 28 [22]:

$$Sh = \frac{2}{\sqrt{\pi}} \mathrm{Re}^{\frac{1}{2}} Sc^{\frac{1}{2}} \left(\frac{2\Delta R}{R\theta} \right)^{\frac{1}{2}} \tag{28}$$

with:

$$Sh = \frac{2k_g \Delta R}{D}$$

$$\text{Re} = \frac{2\rho\, v_{air} \Delta R}{\eta} \tag{29}$$

$$Sc = \frac{\eta}{\rho\, D}$$

The mass transfer coefficient changes, because the thickness of the boundary layer changes as a function of the distance covered by the air flow. This thickness will increase until it reaches the size of the air gap underneath the clothing, after which it becomes constant. Thus Sherwood will become constant when:

$$\theta > \frac{4v_{air} R \Delta R^2}{\pi\, D} \tag{30}$$

For large distances:

$$\text{Sh} = 1 \tag{31}$$

The relations 28 and 29 give the mass transfer coefficient (the Sherwood number) as a function of the air velocity underneath the clothing (the Reynolds number). These equations, together with the flow distribution model as described earlier, can be used to calculate the concentration distribution underneath the clothing. From this concentration distribution, both a dosage and a deposition distribution (the total amount of deposited mass per unit area) can be derived (equations 32 and 33):

$$Dosage = Ct = \int_0^t C dt \tag{32}$$

$$Deposition = M_d = \int_0^t k_g C dt \tag{33}$$

The vapor deposition process is quantitatively expressed by the deposition velocity, v_d, which is the velocity wherewith the vapor is transferred to the skin (equation 34).

$$v_d = \frac{M_d}{Ct} \tag{34}$$

where v_d represents the vapor deposition velocity and m_d the mass deposited per unit area and Ct the dosage of exposure of the chemical agent.

If the mass transfer coefficient, kg, is independent of time, it will be equal to the deposition velocity, vd. This deposition velocity can be measured as a function of the angle around the cylinder and can be compared with the theoretical mass transfer coefficient for validation.

Deposition Distribution Measurements

The test cylinder was placed in a 19 m^3 test chamber in which the air was circulated at an average linear velocity of 1.8 m/s or 5.1 m/s. The cylinder was exposed to methylsalicylate vapor, as a simulant for mustard gas. The methylsalicylate concentration in the test chamber was continuously monitored by GC-FID. The average concentration was 77±4 mg/m^3 and the duration of a test run was 45 minutes, giving an average dosage Ct around 3500 $mg.min/m^3$. Temperatures varied between 29 and 37 °C. The relative humidity was below 40 %.

The two clothing materials were tested simultaneously. By using four spacer rings, two 10 cm bands of clothing material were wrapped around the cylinder. The bands were 8 cm apart. Due to a difference in stiffness material A was positioned at an average distance of 3.2 mm from the cylinder surface and material B at an average distance of 3.6 mm. In this initial study, only the outer layers of air permeable NBC protective suits were used and the adsorptive carbon layers of the suits were not used. Outer layers are responsible for the major part of the air permeability of the material.

To analyze the mass deposition onto the surface, the cylinder surface was covered by an adsorbing material, carbon containing filter paper (Schleicher & Schuell GmbH, 508). The mass deposition, M_d, was determined by taking 18 mm diameter circular samples from the adsorbing material, followed by extraction with carbon disulphide and GC-FID analysis. The measured deposition velocity was calculated with equation 14. The samples were taken at different angles from the stagnation point (angle = 0 degrees). The recovery of methylsalicylate is more than 90 % for mass depositions in the range of these experiments (50 - 5000 mg/m^2).

Concentration and Deposition Results

Concentration Distribution Around the Cylinder

The air flow distribution was used to calculate the concentration and dosage of agent underneath the clothing. Since the deposition measurements were performed with textile without carbon, the concentration calculations were not only performed for the material with carbon, but also for material without carbon. In figure 17 the calculated results of the concentration are shown for material without carbon and in figure 18 the concentration distribution is shown for material with a carbon layer.

The concentration distributions in time were used to calculate the dosage distributions. In figure 19 the dosage distribution is shown for material without a carbon layer and in figure 20 the calculated results for a clothing material with a carbon layer are shown.

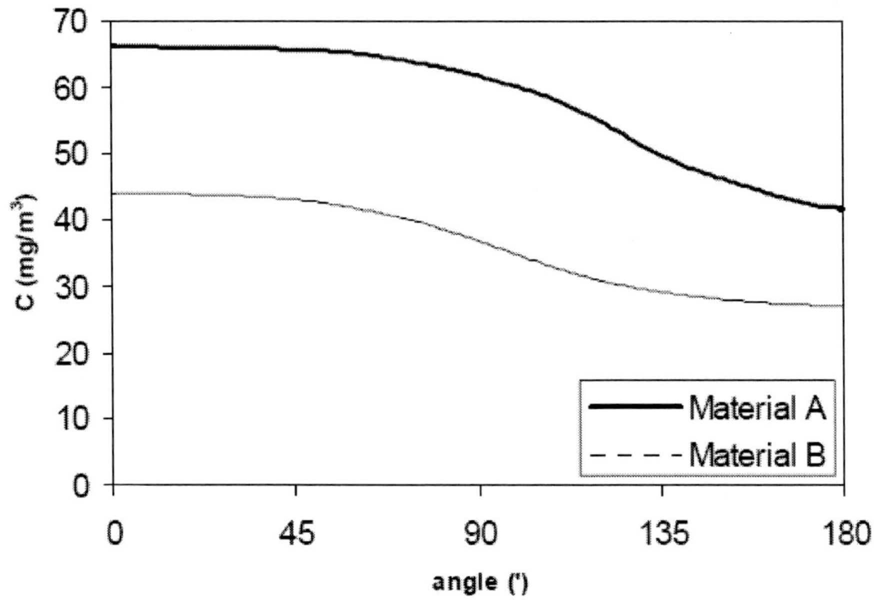

Figure 17. The concentration distribution of vapor underneath clothing around a body part as a function of the angle for clothing material without a carbon layer.

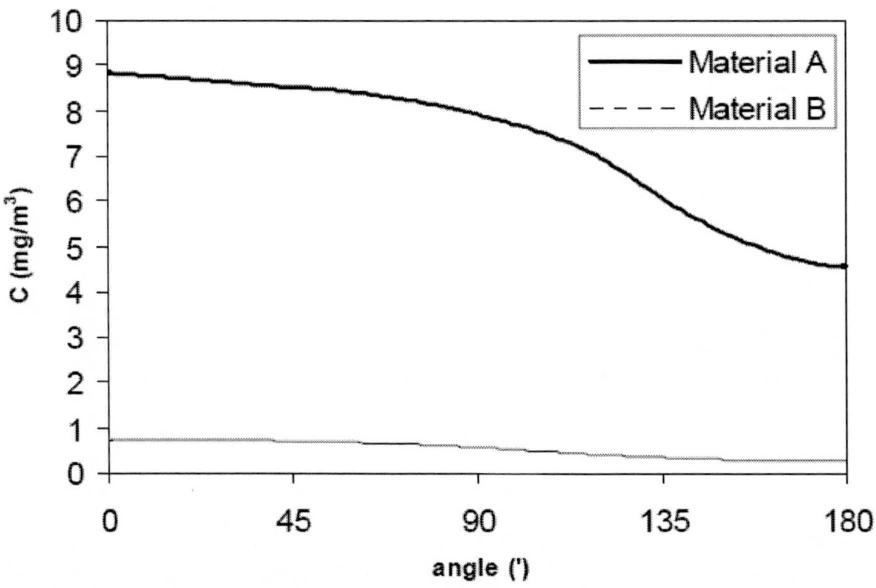

Figure 18. The concentration distribution of vapor underneath clothing around a body part as a function of the angle for clothing material with a carbon layer.

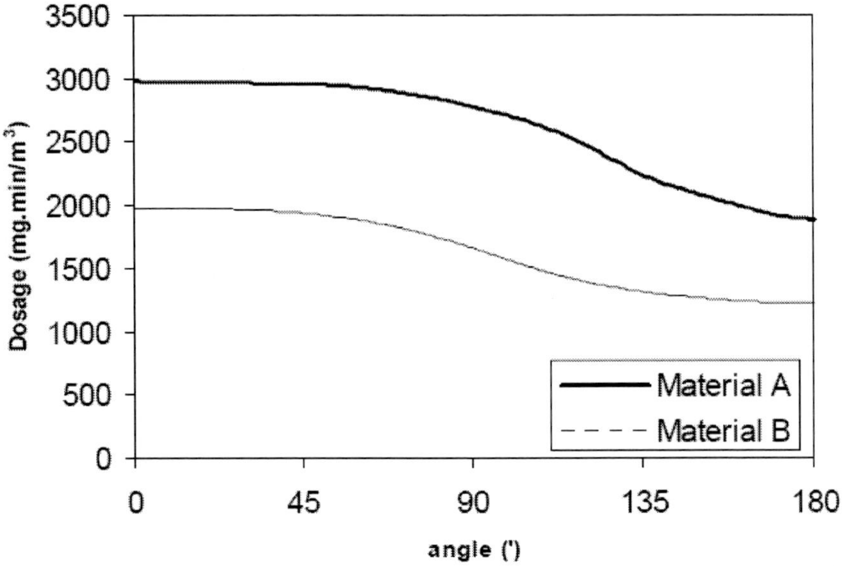

Figure 19. The dosage distribution of vapor underneath clothing around a body part as a function of the angle for clothing material without a carbon layer.

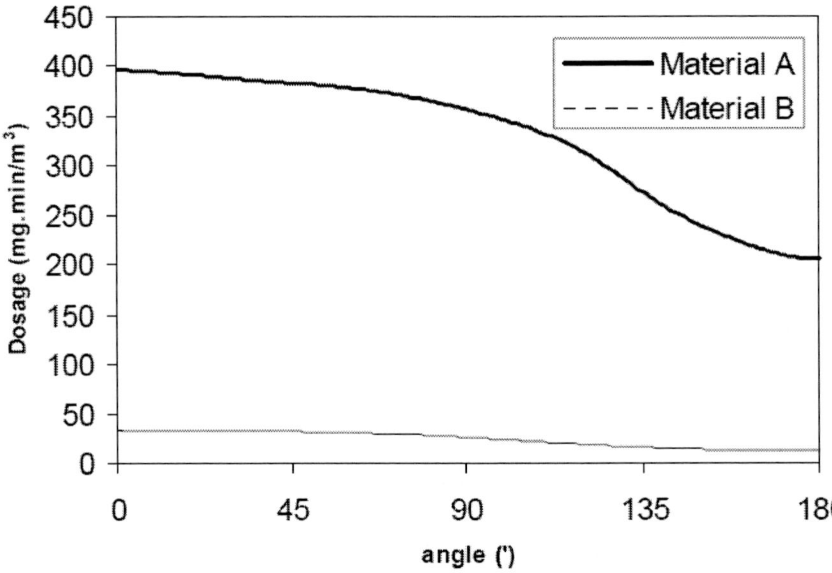

Figure 20. The dosage distribution of vapor underneath clothing around a body part as a function of the angle for clothing material with a carbon layer.

All concentration and dosage distributions show a decrease as a function of the angle. Because the contaminated air will penetrate the clothing at the front, the concentration will be the highest. While flowing in the air gap to the back side of the cylinder, the agent can be adsorbed at the surface of the cylinder or at the carbon of the clothing (if present). This will decrease the concentration. As expected, addition of carbon will result in a lower agent concentration in the air gap. Interesting result is that the absolute factor of this decrease is different for the different materials. Material A (high air permeable) shows a decrease of approximately 7.5 and material B (low air permeable) shows a decrease of circa 40. This effect is studied in more detail in the optimization paragraph.

Deposition Distribution on the Cylinder

In the calculations a constant distance between the clothing and the cylinder wall is assumed, which does not necessarily has to be the case in reality. To calculate the deposition velocity underneath the clothing from the experiments, it was assumed that the dosage underneath the clothing was equal to the dosage outside the clothing. In reality, this dosage will be somewhat lower which will result in a higher measured deposition velocity, and will give an even better match between calculations and experimental values. The measured deposition velocities and the calculated mass transfer coefficients are shown in figure 21.

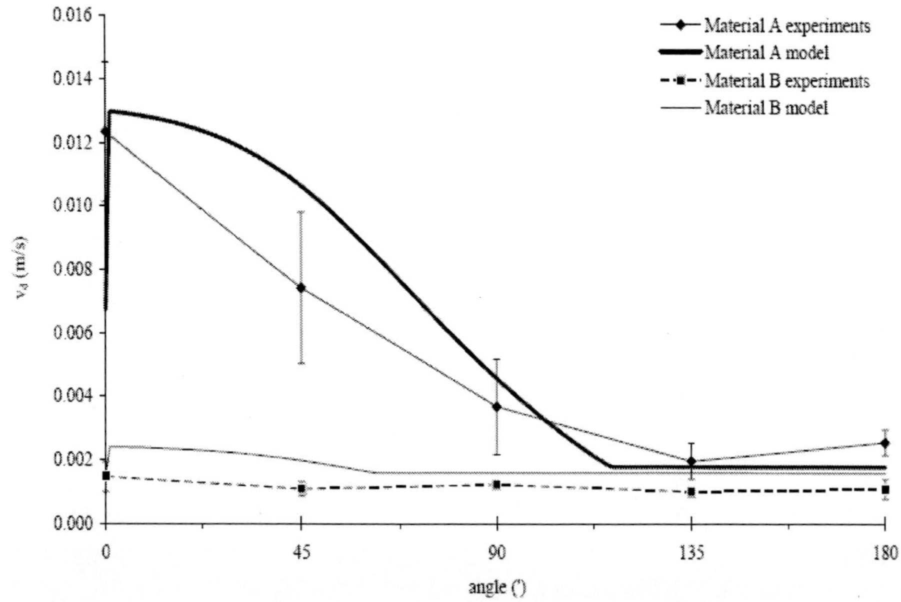

Figure 21. Deposition velocities as a function of the angle around the cylinder. Measurements (interrupted line) calculations (uninterrupted line).The total length of the error bars is two times the standard deviation.

The concentration distribution and the calculated mass transfer coefficients were used to calculate the mass deposition on the surface. In figure 22 the results of these calculations are shown together with the measured mass deposition distribution. The agreement between the measurements and the calculations is reasonably well. Especially the material with high air permeability shows good agreement.

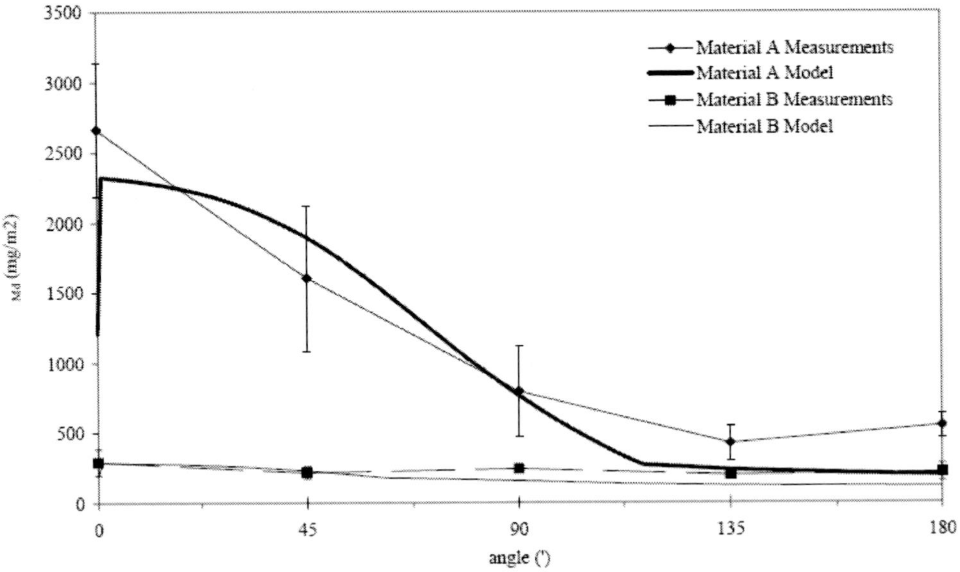

Figure 22. The mass deposition onto the surface of the cylinder as a function of the angle.

When the air permeability of the material increases, the dosage underneath the clothing also increases. This will effect the final mass deposition onto the skin.

Higher air permeability of the clothing results in a higher mass deposition of agent onto the skin (surface of the cylinder). The deposition onto the skin at the front of the cylinder is higher than the deposition at the back side of the cylinder.

Discussion

To find the optimum balance between comfort and protection, still several models have to be developed. The degree of comfort is dependent on the skin and core temperature and the moisture concentration. These parameters have not been modeled yet. However, to give an indication of the position of the optimum, the air velocity underneath the clothing can be taken as a degree of comfort. The total deposited agent mass onto the surface can be taken as a value of the protective behavior. The effect of the air permeability of the clothing onto these two parameters was studied to get some insights into the location of the optimum balance between comfort and protection.

For the velocity through the clothing, the average inward velocity through the clothing was taken. Thus when figure 14 is analyzed, the average value between 0 and 45 degrees is taken as velocity through the clothing. In [6] it was shown that this average velocity is for certain conditions almost equal to the velocity, following from empirical formula 35 [23]:

$$v_{mat} = K_{emp} \Gamma v_{wind}^2 \tag{35}$$

with K_{emp} as an empirical constant. The average velocity underneath the clothing was also taken as parameter. The average velocity through and underneath the clothing, as a function of the air permeability of the clothing, are shown in figure 23.

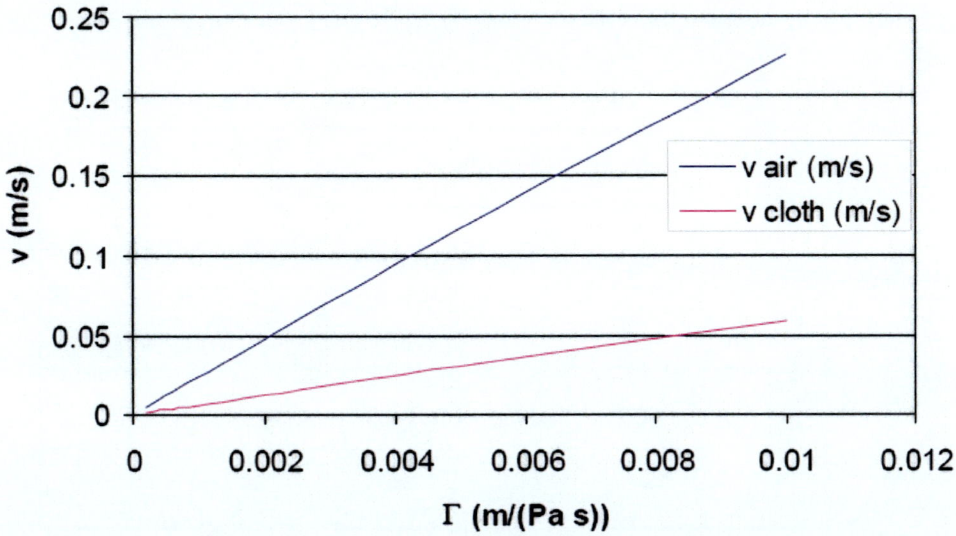

Figure 23. The average velocity through and underneath the clothing, as a function of the air permeability of the clothing.

As expected, both the average velocity through the clothing and underneath the clothing increase linearly with the air permeability of the clothing. Thus the comfort increases with increasing air permeability of the clothing.

The total mass deposition follows from the integral of for instance figure 22. This integral will be the total deposited mass around the cylinder. To extrapolate to a whole person, it was assumed that the total body surface of a person, A_{person}, is 2 m^2. Thus the total deposited mass follows from:

$$M_{total} = \frac{A_{person}}{2\pi} \int_0^{2\pi} M_d d\theta \tag{36}$$

Note that the predicted total mass deposition does not take into account any leakages from for instance closures. In reality this factor can be the major factor to determine the concentration underneath the clothing and thus the total deposition onto the skin. Thus the calculated values do not resemble any values of existing clothing systems.

The total amount of agent, which comes in contact with the person (the total challenge mass) follows from:

$$M_{total,challange} = C_d v_{mat} A_{person} \tag{37}$$

A virtual protection factor, Pf, can be defined as:

$$Pf = \frac{M_{total,challange}}{M_{total}} = \frac{2\pi C_d v_{mat}}{\int_0^{2\pi} M_d d\theta} \quad (38)$$

In figure 24 the calculated protection factor is plotted as a function of the air permeability of the clothing.

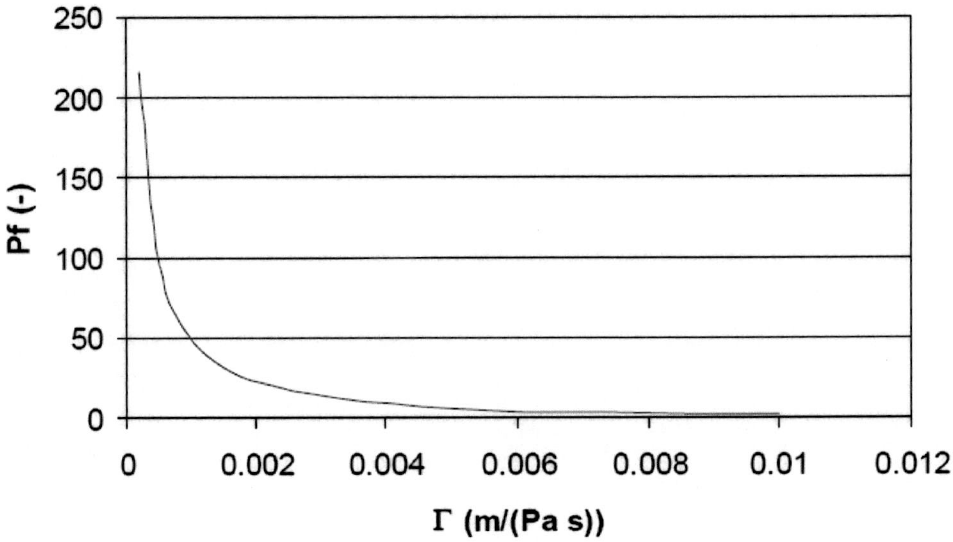

Figure 24. The calculated protection factor as a function of the air permeability of the clothing.

The protection factor of the clothing decreases as a function of the air permeability of the clothing.

When the velocity through or underneath the clothing is taken as measure for comfort, higher comfort can be created with a clothing type with higher air permeability. At the same time higher air permeability will lead to lower protection factors. Thus the optimum balance between comfort and protection will probably be the highest tolerable amount of mass deposition (the lowest tolerable protection factor). This is dependent on the expected scenario (the challenge concentration and the wind velocity).

As discussed, comfort is a combination between optimum core and skin temperature and moisture concentration. Since these values are not known yet, models have to be created to find these values to give better insight into comfort of clothing.

The current models do not yet include leakage of agent through for instance closures. When the models are extended toward a whole system, they can be used to create an optimum design of the suit. Closures and other leakages of agent will influence the concentration underneath the clothing and thus the total deposition onto the skin. Induced air pumping (or the "bellowing effect") will increase the comfort. When these aspects are modeled as well, a better optimum can be found.

CONCLUSIONS

To identify the most adequate type of CBRN-protective clothing, a compromise between comfort and protection has to be found. When the protection of CBRN-clothing is high, usually the comfort is low because of the heat stress. And visa versa, when the comfort is high (low heat stress) the protection tends to be low.

As part of this optimization, a modeling tool is being developed, which can be used to optimize the physiological burden effects and protective performance of CBRN-clothing. The tool can be used to find the optimal balance between parameters like temperature, wind velocity, water vapor concentration, protective properties of the CBRN-clothing material and chemical warfare agent concentration and the resulting deposition onto the skin. Three size scales were identified in the protective behavior of a NBC-suit: the material scale, the human limb scale and the whole system scale. Though not all important processes have been modeled yet, several aspects of the protective behavior of permeable NBC-protective textile were modeled already.

On the material scale level a semi-empirical model was deduced and validated that describes the breakthrough concentration of chemical vapor through beat-type NBC-protective clothing material. The model describes the influence of the material properties, the airflow through the material, the challenge concentration, and the type of agent on the chemical barrier properties of the fabric.

A model which describes the pressure distribution and the air velocity distribution through and underneath clothing around a cylinder was developed. The model was compared with experiments. The model and the experiments show good agreement.

Both the increase in wind velocity and air permeability will result in a higher air flow underneath the clothing. Lowering the air gap size will result in a lower flow through the clothing. If the air gap size at the front decreases due to the pressure of the wind, the flow through the clothing will also decrease. The flow underneath the clothing will increase at the back side of the body part because of the increasing air gap size at the back.

A model, which describes the mass transfer process under the protective clothing, was developed and compared with experiments. Deposition distributions of vapor onto the surface of a cylinder underneath clothing around a cylinder were measured and calculated. The agreement between the measurements and the calculations is adequate for its application.

The deposition velocity and the mass deposition are at its maximum at the front of the cylinder compared to the wind direction. Higher wind speeds and higher air permeabilities of the clothing will lead to higher depositions.

Parameter study shows that higher comfort can be created with a clothing type with higher air permeability. At the same time higher air permeability will lead to lower protection factors. Thus the optimum balance between comfort and protection will probably be the lowest tolerable protection factor.

ACKNOWLEDGEMENT

The author wishes to thank Fred Lalleman, L. Mota and G.J. Woudenberg for performing the experimental part of the work. Furthermore he wishes to thank the Dutch Ministry of Defence for financing of this project.

SYMBOLS

Roman Alphabet

A_{person}	Total surface of a person	(m^2)
C	Concentration	(kg/m^3)
C_0	Outside concentration	(kg/m^3)
C_{ini}	Initial breakthrough concentration	(kg/m^3)
C_s	Surface concentration	(kg/m^3)
C_{sat}	Saturation concentration	(kg/m^3)
D	Diffusion coefficient	(m^2/s)
d_{mat}	Thickness of clothing material	(m)
E_0	Activation energy	(J/mol)
g	gravity constant	(m/s^2)
K_D	Darcy constant	(-)
k_g	Mass transfer coefficient	(m/s)
L	Length of Cylinder	(m)
Lo	Carbon load	(kg/m^2)
M_d	Mass deposition	(kg/m^2)
P	Local pressure	(Pa)
P_{in}	Local pressure in air gap	(Pa)
P_{out}	Local pressure outside clothing	(Pa)
R	Radius of cylinder	(m)
r_g	location in air gap	(m)
R_{gas}	Gas constant	(J/mol K)
q	Adsorption	(kg/m^3)
q_{max}	Maximal adsorption	(kg/m^3)
r	Radius carbon sphere	(m)
ΔR	Half size of air gap	(m)
R_{cloth}	Radius of clothing cylinder	(m)
R_{gas}	Ideal gas constant	(J/molK)
Re	Reynolds number	(-)
Sc	Schmidt number	(-)
Sh	Sherwood number	(-)
Fo	Fourier number	(-)
t	Time	(s)
t_{50}	50 % breakthrough time	(s)
T	Temperature	(K)

v_0	Wind velocity	(m/s)
v_{air}	Air velocity in air gap	(m/s)
v_d	Deposition velocity	(m/s)
v_{mat}	Air velocity through clothing	(m/s)
v_r	Local air velocity in air gap	(m/s)
x	Distance along surface of cylinder	(m)
z	Axial position in filter	(m)

Greek Alphabet

β	Affinity coefficient	(-)
δ	Boundary thickness	(m)
ε	Porosity of clothing	(-)
η	Dynamic viscosity of air	(Pa s)
χ	Dimensionless distance	(-)
Γ	Air permeability of clothing	(m/(Pa s))
ξ	Dimensionless concentration	(-)
ξ_s	Dimensionless surface concentration	(-)
ρ	Density of air	(kg/m3)
ρ_c	Density of carbon	(kg/m^3)
σ	Steepness of error function	(-)
τ	Dimensionless time	(-)
θ	Angle around cylinder	(°)
ϕ_v	Flow in air gap	(m^3/s)

References

[1] Sobera, MP; Kleijn, CR; Brasser, P; Akker, HEA van den. A Multi-scale Numerical Study of the Flow, Heat, and Mass Transfer in Protective Clothing. In: Bubak M et al. Editors. *ICCS 2004, LNCS 3039*, Berlin Heidelberg: Springer-Verlag; 2004; 637–644.

[2] Barry, J; Hill, R; Brasser, P; Sobera, MP; Kleijn, CR; Gibson P. Computational fluid dynamic modeling of fabric systems for intelligent garment design. *MRS Bulletin*, 2003 28, 568–573.

[3] Sobera, MP; Kleijn, CR; Akker, HEA van den; Brasser, P. Convective heat and mass transfer to a cylinder sheathed by a porous layer. *AIChE J.* 2003 49, 3018–3028.

[4] Gibson, P; Barry, J; Hill, R; Brasser, P; Sobera, MP; Kleijn, CR. Computer modeling of heat and mass transport in protective clothing. In: Pan, N; Gibson, P editors. *Thermal and Moisture Transport in Fibrous Materials, Part 3 Textile-body interactions and modeling issues*, ISBN 1 84569 057 5 ,Woodhead Publishing 2006, Chapter 15

[5] Brasser P. Modeling the Chemical Protective Performance of NBC Clothing Material. *Journal of Occupational and Environmental Hygiene*, 2004 1, 620–628.

[6] Brasser P. Theoretical and Experimental Study of Airflow Through Clothing around Body Parts. *AIChE J.*, 2006 52(11), 3688 – 3695.

[7] Brasser P; Houwelingen, T van. A Theoretical And Experimental Study Of The Vapor Deposition Onto The Surface Of A Dressed Body Part. *AIChE J.*, 2008 54(4), 844 – 848.

[8] Linders, MJG. Breakthrough behavior of carbon columns under dry and humid conditions. In: *Prediction of breakthrough curves of activated carbon based sorption systems*. ISBN: 90-6464-512-4, Delft 1999, 163-188.

[9] Linders, MJG; Mallens, EPJ; Bokhoven, JJGM van; Kapteijn F; Moulijn JA; Breakthrough of Shallow Activated Carbon Beds under Constant and Pulsating Flow. *AIHA J.* 2003 64(2), 173-180.

[10] Wakao, N; Funazkri, T. Effect of fluid dispersion coefficients on particle-to-fluid mass transfer coefficients in packed beds. *Chem. Eng. Sci.* 1978 33, 1375-1384.

[11] Smith, JM; Stammers, E; Janssen, LPBM. *Fysische transportverschijnselen I.* Delft: Delftse Uitgevers Maatschappij 1997.

[12] NATO Army Armaments Group (NAAG) Land Group 7 on NBC Defence (LG/7): *Operational requirements, technical specification and evaluation criteria for NBC protective clothing AEP-38*. Brussel, NATO, 1998 (NATO confidential).

[13] Payne, PR. The theory of fabric porosity as applied to parachutes in incompressible flow. *Aeronautic Quarterly.* 1978, 175.

[14] Hoke, L; Segars, RA; Cohen, S; King, A; Johnson, E. Low speed air-flow characterization of military fabrics. *NATICK/TR-89/013* Massachusetts: United States Army Natick Research, Development and Engineering Center 1988.

[15] Matteson, MJ; Orr, C. *Filtration Principles and Practices*. New York: Marcel Dekker, Inc. 1987.

[16] Fedele, PD; Bergman, W; McCallen, R; Sutton, S. Hydrodynamically induced aerosol transport through clothing. Philadelphia: *American Society for Testing and Materials*. 1987.

[17] Fedele, PD. Model of aerosol protection offered by permeable protective garments. Philadelphia: *American Society for Testing and Materials.* 1992.

[18] Streeter, VL; Wylie EB. *Fluid mechanics; first SI metric edition.* ISBN: 0-07-548015-8. Auckland: McGraw-Hill 1987.

[19] Kind, RJ; Jenkins, JM; Seddigh, F. Experimental investigation of heat transfer through wind- permeable clothing. *Cold regions Science and Technology*, 1991 20, 39.

[20] Kind, RJ; Jenkins, JM; Broughton, CA. Measurements and prediction of wind-induced heat transfer through permeable cold-weather clothing. *Cold regions Science and Technology,* 1995 23, 305.

[21] Bennett, CO; Myers, JE. *Momentum, heat, and mass transfer*, ISBN: 0-07-004671-9, New York :McGraw-Hill Book company, 1982, 3rd edition.

[22] Lyklema, J. Fundamentals of Interface and Colloid Science, Volume 1, Fundamentals, ISBN 0-12-460525-7, Norfolk, UK: Academic Press Ltd. 1991.

[23] Wal, JF van der. Permeabele bescherming kleding, II De penetratie van aerosol door kleding. Rijswijk: TNO-report 1967-22, 1967.

In: Porous Media: Heat and Mass Transfer... ISBN: 978-1-60692-437-2
Editors: J. L. Acosta and A.F. Camacho

Short Communication A

SORET AND DUFOUR EFFECT ON DOUBLE DIFFUSIVE NATURAL CONVECTION IN A WAVY POROUS ENCLOSURE

B. V. R. Kumar[1,2,3], S.Belouettar[2], S. V. S. S. N. V. G. K. Murthy[1], Vivek Sangwan[1], Mohit Nigam[1], Shalini[4], D.A.S.Rees[5] and P.Chandra[1]

[1] Indian Institute of Technology Kanpur, UP-208016, India
[2] CRP Henri Tudor, LTI, 29 Ave. J.F.K, L-1855, Luxembourg
[3] ITWM, Fraunhofer Institute I, Kaisersluatern, Germany
[4] INRIA, Rocquencourt, BP 105, 78153, Le Chesnay, France
University of Bath, Bath, BA2 7AY, UK [5]

ABSTRACT

In this study the influence of Soret and Dufour effects on the double diffusive natural convection induced by an heated isothermal wavy vertical surface in a fluid saturated porous enclosure under Darcian assumptions has been analysed. The mathematical model has been solved numerically by finite element method and the simulations are carried out for various values of parameters such as D_f (Dufour Number), S_r (Soret Number), *Le* (Lewis Number), *B* (Buoyancy Number) and *N* (Number of waves per unit length) at small values of *Ra* (Rayleigh Number).

NOMENCLATURE

A	amplitude of the wavy wall
g	gravitational acceleration
k	thermal conductivity
K	permeability of the medium

K_T	thermal diffusion ratio
C_s	concentration susceptibility
C_p	specific heat at constant pressure
L	the length or the mean width of the porous cavity
n	outward drawn unit normal to the wavy surface
N	number of waves considered per unit length
Nu	Nusselt number
Sh	Sherwood Number
Q	cumulative heat flux
Ra	Rayleigh number, $(Kg\beta\Delta tL)/\nu\alpha$ based on the dimension of porous cavity
Le	Lewis Number
B	Buoyancy Ratio
D	Mass Diffusivity
QH_X	Cumulative Heat flux
QM_X	Cumulative Mass flux
$S(\xi)$	arc length of the wavy wall
t	temperature
T	non-dimensional temperature
u,v	velocity components in x and y directions
U,V	non-dimensional velocity component in X and Y directions
V_c	convective velocity, $(g\beta K\Delta t)/\nu$
w	weight function used in finite element formulation
x,y	cartesian co-ordinates
X,Y	non-dimensional cartesian co-ordinates

Greek Symbols

α	thermal diffusivity constant
β	thermal expansion constant
ρ	fluid density
ψ	non-dimensional stream function
ν	fluid kinematic viscosity
ξ	arc length variable
Ω	the domain considered in the problem
Γ	the boundary of the domain

Subscripts

f	for fluid
w	evaluated at the wall temperature

a evaluated at the ambient medium

1. Introduction

Study of coupled heat and mass transfer by natural convection in a fluid saturated porous medium has attracted considerable attention in a wide range of fields like oceanography, astrophysics, geology, nuclear engineering, chemical processes etc. It has gained lot significance due to it direct relevance in applications such as contaminant transport in ground water, nuclear waste management, separation process in chemical engineering, reservoirs of crude oil, geo-thermal reservoirs etc. A number of investigations have been carried out on Double Diffusive (DD) free convection process in a fluid saturated porous medium under various assumptions [1-6].

Diffusion of matter caused by temperature gradients (Soret Effect) and diffusion of heat caused by concentration gradients (Dufour Effect) become significant when temperature and concentration gradients are very large. Generally these effects are considered as second order phenomenon. Eckert and Drake [7], Zimmerman and Muller [8], Hurle and Jackerman [9], Bergman and Srinivasan [10], Weaver and Viskanta [11], Benano-Melly et al [12] etc., have investigated the importance of these effects on the convective heat transfer in fluids. However, regarding their influence on the DD free convection in a porous media not much has been reported in the literature. Anghel et al [13], Postelnicu [14], Sovran et al [15], Partha et al [16] etc. have investigated analytically the influence of either one or both of these effects on free convection flow induced by an isothermal vertical surface in an electrically conducting Darcian fluid saturated porous medium under boundary layer assumptions.

Attachment of baffles, fins or other suitable protrusion to the hot surface of fluid saturated porous enclosure can affect convection process in the system and the process is used in several engineering applications related to building technology, cold storage units etc. Semi-conductor devices are intentionally roughened to alter their heat transfer capabilities. Riley [17], Rees and Pop [18-19], Murthy et al [20], Rathish Kumar [21-22] etc., have attempted to analyze natural convection heat transfer in porous media approximating the surface undulations by periodic functions.

In this study we consider a fluid saturated wavy porous enclosure under Darcian assumptions without any boundary layer assumptions. The coupled nonlinear partial differential equations, modelling the influence of Soret and Dufour effect on DD natural convection process in the vertical wavy enclosure, are solved numerically by Galerkin finite element method. Simulations are carried out for various values of D_f, S_r, Le, N and B and the results are depicted through streamlines, isotherms, iso-concentration contours and *xy*-plots.

2. Mathematical Model

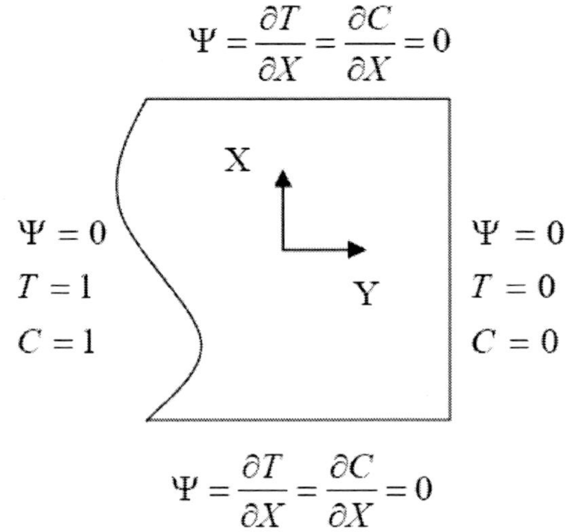

Figure 1.

Consider a fluid saturated isotropic homogenous porous enclosure (Figure 1) with a wavy left vertical surface at a constant temperature t_w and a constant wall concentration c_w. The right vertical wall is maintained at the ambient temperature t_a ($< t_w$) and at the ambient concentration c_a ($< c_w$). The fluid is assumed to satisfy the Boussinesq approximation and the flow is assumed to follow the Darcy law. Following Lai and Kulacki [10], Angirasa et al [11], Postelnicu [8], Partha [12] etc. the equations governing the heat and mass transfer process in the presence of Soret and Dufour effects, in non-dimensional form are written as follows:

$$\left(\frac{\partial^2\psi}{\partial X^2}+\frac{\partial^2\psi}{\partial Y^2}\right)=\frac{\partial T}{\partial Y}+B\frac{\partial C}{\partial Y} \tag{1}$$

$$\frac{\partial\psi}{\partial Y}\frac{\partial T}{\partial X}-\frac{\partial\psi}{\partial X}\frac{\partial T}{\partial Y}=\frac{1}{Ra}\nabla^2 T+D_f\nabla^2 C \tag{2}$$

$$\frac{\partial\psi}{\partial Y}\frac{\partial C}{\partial X}-\frac{\partial\psi}{\partial X}\frac{\partial C}{\partial Y}=\frac{1}{Ra\,Le}\nabla^2 C+S_r\nabla^2 T \tag{3}$$

with the boundary conditions:

$$\begin{aligned} &T=C=1, \psi=0 \quad on\, Y=a\ \sin(N\pi X-\phi);\\ &T=C=\psi=0 \quad on\, Y=1\\ &\frac{\partial T}{\partial X}=\frac{\partial C}{\partial X}=\psi=0 \quad on\ X=0,1 \end{aligned} \tag{4}$$

where the non-dimensional variables and the parameters are defined as follows:

$$\left.\begin{array}{l} X=\dfrac{x}{L},\ Y=\dfrac{y}{L},\ T=\dfrac{t-t_a}{t_w-t_a},\ C=\dfrac{c\ -\ c_a}{c_w-c_a}, Ra=\dfrac{Kg\beta\Delta tL}{\nu\alpha}, \\ U=\dfrac{u}{V_c},\ V=\dfrac{\mathrm{v}}{V_c}\ \text{where}\ V_c=\dfrac{g\beta K\Delta t}{\nu},\ \dfrac{\partial\psi}{\partial Y}=U,\ \dfrac{\partial\psi}{\partial X}=-V. \\ B=\dfrac{\beta_c\left(c_w-c_a\right)}{\beta_t\left(t_w-t_a\right)}, Le=\dfrac{\alpha}{D}, D_f=\dfrac{DK_T\left(c_w-c_a\right)}{C_sC_p\left(t_w-t_a\right)}, S_r=\dfrac{DK_T\left(t_w-t_a\right)}{C_sC_p\left(c_w-c_a\right)} \end{array}\right\} \quad (5)$$

The cumulative global heat flux and the cumulative global mass flux are computed by the formula:

$$QH_X=\int_0^X\left(-\frac{\partial T}{\partial n}\right)\Big|_{Y=a\sin(N\pi x-\phi)}\frac{dS(\xi)}{d\xi}d\xi,\ QM_X=\int_0^X\left(-\frac{\partial C}{\partial n}\right)\Big|_{Y=a\sin(N\pi x-\phi)}\frac{dS(\xi)}{d\xi}d\xi \quad (6)$$

where 'n' is the outward normal to the wavy surface and $S(\xi)$ is the arc-length along the surface. $X=1$ i.e. the upper limit gives the global heat flux or the Nusselt number (Nu) and global mass flux or the Sherwood Number (Sh). The mathematical model is solved by finite element method.

3. Finite Element Formulation

Let Ω denote the domain of interest and Γ be the boundary of the domain. The discretized representation of Ω is given by $\overline{\Omega}=\bigcup_{e=1}^{NEL}\Omega^e$ where Ω^e denotes a typical bilinear element of the discretized domain and NEL is total number of such elements. The discretized elements are fully disjoint i.e. $\bigcap_e\Omega^e=\{\}$. The discretized representation of the field variables ψ, T, C on a typical bilinear element Ω^e is:

$$\psi=\sum_{i=1}^{4}\psi_i^eN_i^e,\ T=\sum_{i=1}^{4}T_i^eN_i^e,\ C=\sum_{i=1}^{4}C_i^eN_i^e \quad (7)$$

where N_i^e denotes the standard bi-linear interpolation function on a typical element Ω^e.

Now consider the Galerkin Weighted Residual form of the governing equations (1)-(3) on a Ω^e:

$$\int_{\Omega^e} \left(\frac{\partial^2 \psi}{\partial X^2} + \frac{\partial^2 \psi}{\partial Y^2} - \frac{\partial T}{\partial Y} - B\frac{\partial C}{\partial Y}\right) W_i \, d\Omega^e = 0 \tag{8}$$

$$\int_{\Omega^e} \left\{\left(\frac{\partial \psi}{\partial Y}\frac{\partial T}{\partial X} - \frac{\partial \psi}{\partial X}\frac{\partial T}{\partial Y}\right) - \frac{1}{Ra}\left(\frac{\partial^2 T}{\partial X^2} + \frac{\partial^2 T}{\partial Y^2}\right) - D_f\left(\frac{\partial^2 C}{\partial X^2} + \frac{\partial^2 C}{\partial Y^2}\right)\right\} W_i \, d\Omega^e = 0 \tag{9}$$

$$\int_{\Omega^e} \left\{\left(\frac{\partial \psi}{\partial Y}\frac{\partial C}{\partial X} - \frac{\partial \psi}{\partial X}\frac{\partial C}{\partial Y}\right) - \frac{1}{RaLe}\left(\frac{\partial^2 C}{\partial X^2} + \frac{\partial^2 C}{\partial Y^2}\right) - S_r\left(\frac{\partial^2 T}{\partial X^2} + \frac{\partial^2 T}{\partial Y^2}\right)\right\} W_i \, d\Omega^e = 0 \tag{10}$$

On rewriting the equations (8)-(10) in the weak form and on introducing the element level discretized representation for the field variables i.e. (7) into these modified equations one would arrive at the following element matrix equation:

$$M^e a^e = f^e \tag{11}$$

where

$$M^e_{ij} = \begin{bmatrix} A^{11}_{ij} & A^{12}_{ij} & A^{13}_{ij} \\ 0 & A^{22}_{ijk} & A^{23}_{ij} \\ 0 & 0 & A^{33}_{ijk} \end{bmatrix} \tag{12}$$

$$a^e_j = \begin{bmatrix} \psi^e_j & T^e_j & C^e_j \end{bmatrix}^T \tag{13}$$

$$f^e_i = \begin{bmatrix} f^1_i & f^2_i & f^3_i \end{bmatrix}^T \tag{14}$$

here,

$$A^{11}_{ij} = \int_{\Omega^e} -\left(\frac{\partial N^e_i}{\partial X}\frac{\partial N^e_j}{\partial X} + \frac{\partial N^e_i}{\partial Y}\frac{\partial N^e_j}{\partial Y}\right) d\Omega^e \tag{15}$$

$$A^{12}_{ij} = \int_{\Omega^e} -\left(N^e_i \frac{\partial N^e_j}{\partial Y}\right) d\Omega^e \tag{16}$$

$$A_{ij}^{13} = \int_{\Omega^e} -(BN_i^e \frac{\partial N_j^e}{\partial Y}) d\Omega^e \tag{17}$$

$$A_{ijk}^{22} = \int_{\Omega^e} (\{\sum_{k=1}^{4} (\frac{\partial N_k^e}{\partial Y}\frac{\partial N_j^e}{\partial X} - \frac{\partial N_k^e}{\partial X}\frac{\partial N_j^e}{\partial Y})\psi_k^e\} N_i^e)\, d\Omega^e$$

$$+ \int_{\Omega^e} (\frac{1}{Ra}(\frac{\partial N_i^e}{\partial X}\frac{\partial N_j^e}{\partial X} + \frac{\partial N_i^e}{\partial Y}\frac{\partial N_j^e}{\partial Y}))\, d\Omega^e \tag{18}$$

$$A_{ijk}^{23} = \int_{\Omega^e} D_f (\frac{\partial N_i^e}{\partial X}\frac{\partial N_j^e}{\partial X} + \frac{\partial N_i^e}{\partial Y}\frac{\partial N_j^e}{\partial Y})\, d\Omega^e \tag{19}$$

$$A_{ijk}^{32} = \int_{\Omega^e} -S_r (\frac{\partial N_i^e}{\partial X}\frac{\partial N_j^e}{\partial X} + \frac{\partial N_i^e}{\partial Y}\frac{\partial N_j^e}{\partial Y})\, d\Omega^e \tag{20}$$

$$A_{ijk}^{33} = \int_{\Omega^e} (\{\sum_{k=1}^{4} (\frac{\partial N_k^e}{\partial Y}\frac{\partial N_j^e}{\partial X} - \frac{\partial N_k^e}{\partial X}\frac{\partial N_j^e}{\partial Y})\psi_k^e\} N_i^e)\, d\Omega^e$$

$$- \int_{\Omega^e} (\frac{1}{RaLe}(\frac{\partial N_i^e}{\partial X}\frac{\partial N_j^e}{\partial X} + \frac{\partial N_i^e}{\partial Y}\frac{\partial N_j^e}{\partial Y}))\, d\Omega^e \tag{21}$$

In view, of the numerical boundary conditions and their subsequent treatment in the solution process one may take without any loss of generality the r.h.s vector to be a zero i.e.

$$\left(f_i^1, f_i^2, f_i^3\right) = (0,0,0) \tag{22}$$

The non-linear global system obtained by assembling the local elemental matrix systems (12) is solved iteratively by out of core frontal method for non-linear symmetrical systems to an accuracy of $\varepsilon = 5\times10^{-5}$ on the relative error of nodal field variables from successive iteration i.e. $|\xi_i^{n+1} - \xi_i^n| \leq \varepsilon$ where $\xi_i^n = \psi_i^n (or) T_i^n (or) C_i^n$. Here the superscript n refers to the iteration level and .i refers to the nodal point index.

4. Results and Discussion

The various parameters that govern the double diffusive natural convection under the influence of Soret and Dufour effects in a vertical square porous enclosure, with wavy left wall, are *B* (buoyancy ratio), *Le* (Lewis Number), *Ra* (Rayleigh Number), D_f (Dufour Number), S_r (Soret Number), Number of waves per unit length (*N*), wave amplitude (*a*) and wave phase (ϕ). Numerical simulations have been made for a wide range of these parameters to analyse the influence of Soret and Dufour effects on combined heat and mass transfer due to natural convection in a vertical wavy porous enclosure. For now, as per the literature [13-18] we take Ra to be $o(1)$. To begin with a grid selection test has been carried out. Five different grid systems consisting of $21\times21, 31\times31, 41\times41, 51\times51, 61\times61$ grid points have been considered. On these grid systems simulations have been carried out for various combinations of parameters and found that in all the cases the grid system 41×41 is adequate. Even as one moves away from the 31×31 to higher grid systems in all most all cases only a small change less than 1% in the field variable is noticed. As a sample in Table 1 we provide the comparison of Nusselt Number values calculated on different grid systems for a set of parameters. As a matter of fact grid validation tests have been carried out in even under different physical situations too [23-24].

Table 1. Nusselt Number Values on different grid system for $Ra = 100$, $a = 0.5$, $N = 1$, $Le = 1$, $S_r = D_f = B = 0$

Grid Size	Nusselt Number
21X21	1.724276
31X31	1.790972
41X41	1.818385
51X51	1.831584
61X61	1.832649

Table 2. Comparison of Nusselt Number values with those from literature for $Ra = 100$, $a = 0.0$, $N = 1$, $Le = 1$, $S_r = D_f = B = 0$

Rayliegh Number	Wolker & Homsy [28]	Trevisan & Bejan [27]	Beckermann Et al [25]	Shiralkar Et al [26]	Present study
50	1.98	2.02	1.981	-	1.966
100	3.09	3.27	3.113	3.118	3.028
200	4.89	5.61	5.038	4.976	5.448
500	8.66	-	9.308	8.944	8.348

Further we also validate the code on the chosen 41×41 grid system with the results as available from the literature. Table 2 presents one such comparison of Nusselt Number values.

To begin with, we look at the influence of Soret effect fixing D_f = 0.5, *Le, N, B* = 1, *a* = 0.1. In figures 2-3 cumulative heat flux (QH_X) and mass flux (QM_X) along the wavy vertical wall has been plotted for $0.1 \le S_r \le 4.0$. While QH_X increases with increasing soret effect QM_X is seen to decrease. QH_X plots project the presence thermal boundary layer near the hot wavy wall. These thermal boundary layers get increasingly sharp with increasing values of S_r. QM_X plots also project the presence of concentration boundary layers but unlike to the thermal gradients, here the mass gradients are seen to smaller with increasing S_r thereby leading to the loss in the sharpness in concentration boundary layers. Also the plots project that while the local heat fluxes tend to get marginal, especially after nearly half the distance from the lead edge of the wavy wall, the concentration fluxes tend remain relatively significant even far away from the leading edge of the wavy wall. In Figure 4 variation in *Nu* and *Sh* with increasing values of S_r is presented. Clearly while *Nu* increases with increasing values of S_r, *Sh* are seen to decrease. In order to get a deeper insight it is better to trace the temperature and concentration variable fields. In Figures 5-7 we present the streamlines, isotherms and iso-concentration contours for the current set of parameter values. From the streamlines we notice that with increasing S_r the uni-cellular flow circulation pattern changes to a multi-cellular pattern. The eye of the primary circulation also drifts from the lower left corner towards the top wall with a marked change in the flow orientation. From the isotherms and iso-concentration contours one can find that with increasing values of S_r, while the iso-concentration contours lead to the formation of two localized patterns, one along the wavy wall and the other near the top right corner of the enclosure, the isotherms spread shifts from the bottom-top orientation to a completely diagonal path starting from the left bottom corner of the wavy isothermal wall. The isotherms clearly depict the presence of increasingly sharper thermal boundary layers. The variation in the isotherm line alignment depicts the situation of increased heat flux into the domain. The Iso-concentration contours depict the situation of reduced concentration flux into the domain and clearly justify the observed reduction in the *Sh* with increasing values of S_r. Increase in Soret effect favors a quick spread and thus stabilization in masses. In effect with the additional diffusion of matter, due to temperature gradients in the domain, there is an increased heat flux along the hot wavy wall. Contour plots clearly depict the sensitivity of the field variables are significantly influenced by the Soret effect.

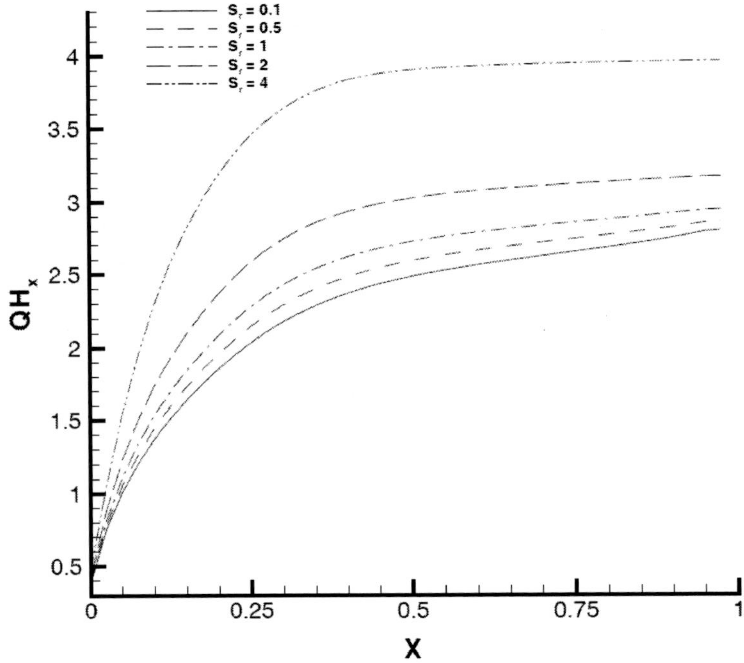

Figure 2. Cumulative heat flux (QH_X) along the wavy wall for different values of S_r.

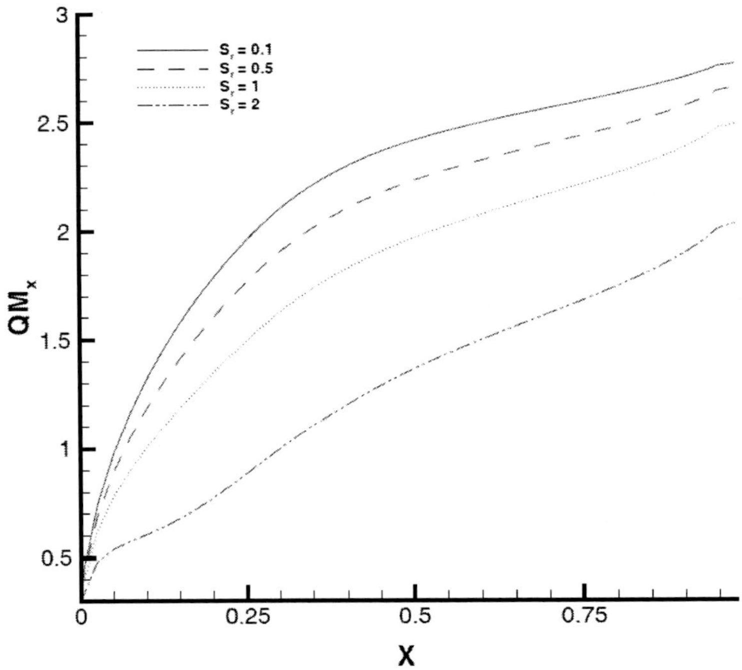

Figure 3. Cumulative mass flux (QM_X) along the wavy wall for different values of S_r.

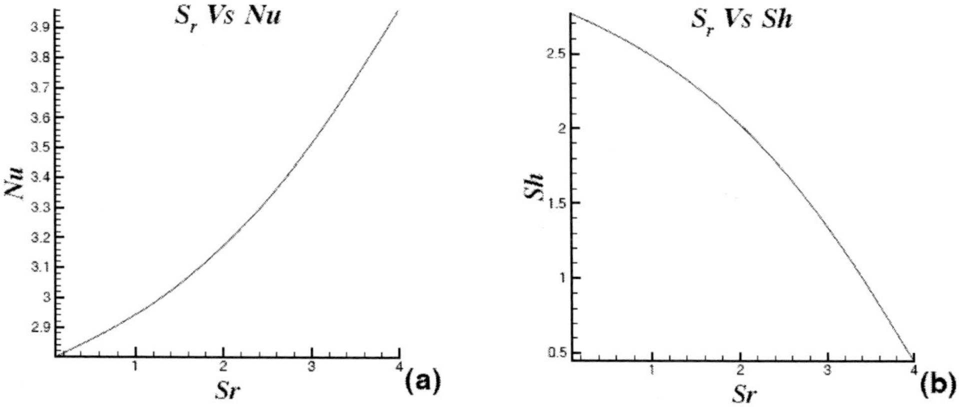

Figure 4. Influence of Soret effect on Global heat flux (*Nu*) and Global Mass flux (*Sh*).

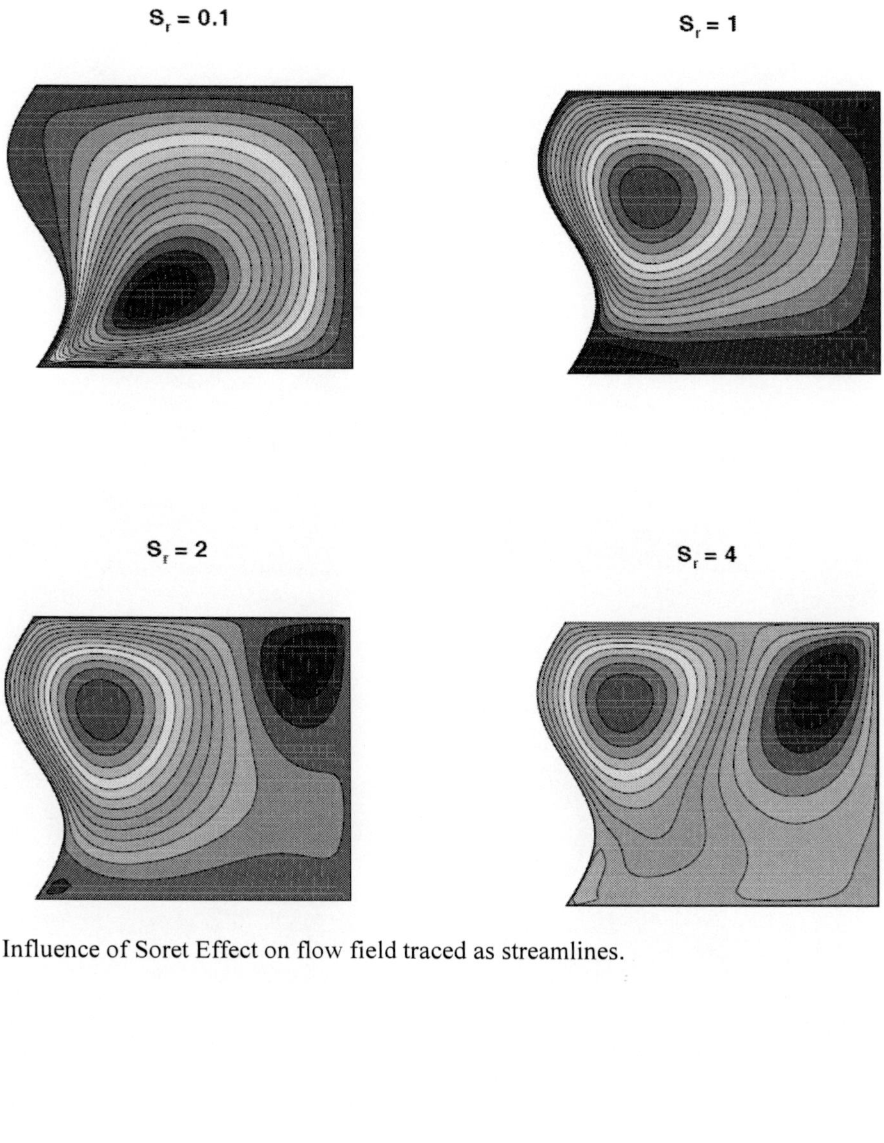

Figure 5. Influence of Soret Effect on flow field traced as streamlines.

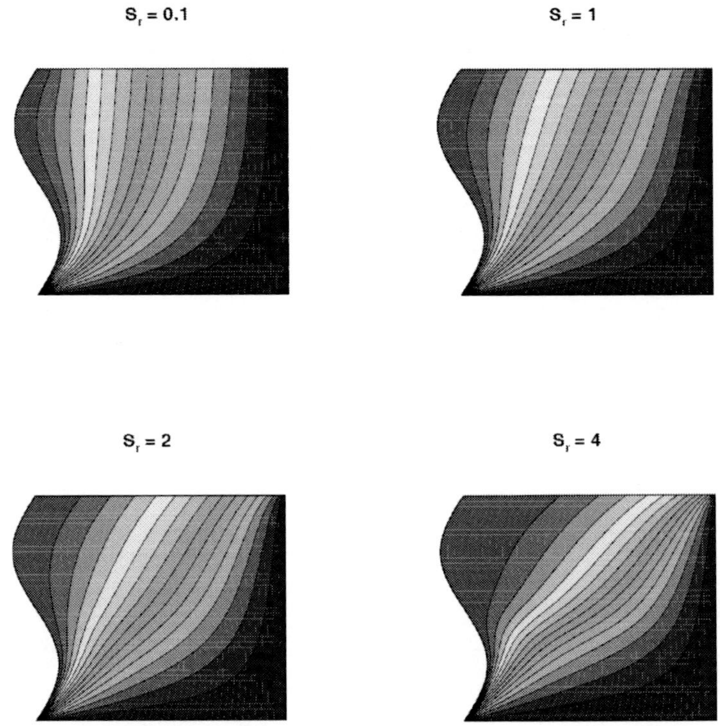

Figure 6. Influence of Soret Effect on Temperature field traced as Isotherms.

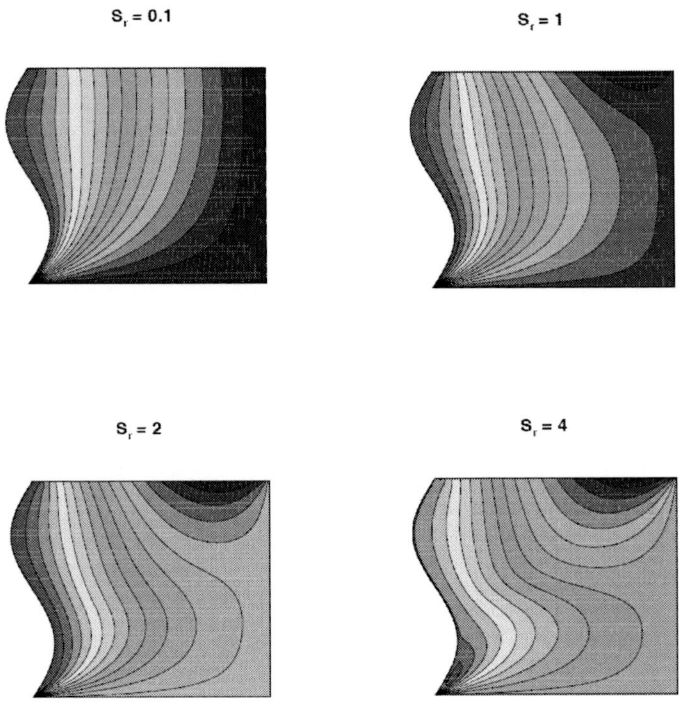

Figure 7. Influence of Soret Effect on Concentration field traced as Iso-Concentration contours.

Next, we look at the influence of Dufour Number (D_f) fixing S_r = 0.5, *Le, N, B* = 1, *a* = 0.1. In the Figure 8 cumulative heat fluxes (QH_X) along the wavy vertical wall has been plotted for $0.1 \le D_r \le 2.0$. QH_X is seen to increase with increasing D_f. The plots in the Figure 9 illustrate the variation of *Nu* and *Sh* with D_f is presented. While there is a slight increase in *Nu* values, the variation in the *Sh* value is marginal. Streamlines, isotherms and iso-concentration contours for D_f = 0.1, 8.0 are presented in Figure 10. The variation in isotherm alignment clearly depict that the temperature field is sensitive to D_f magnitude. The diagonal shift observed in the isotherm pattern with increasing values of D_f supports the observed increase in the heat fluxes along the wavy wall. However, other field variables remain nearly unaffected. Hence the additional diffusion of heat brought in by the concentration gradients primarily affects the temperature field leaving other fields nearly unaltered.

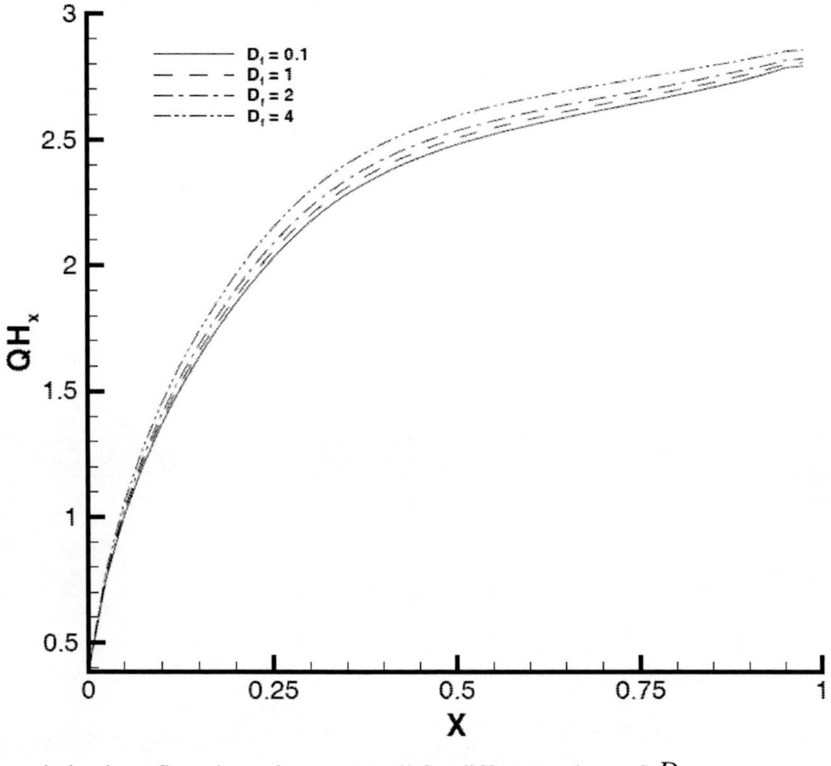

Figure 8. Cumulative heat flux along the wavy wall for different values of D_f.

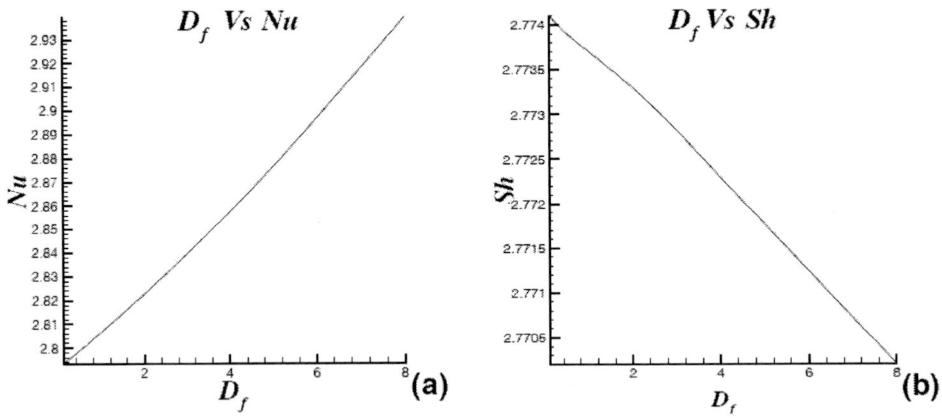

Figure 9. Influence of Dufour Effect on Global Heat Flux (Nu) and Global Mass Flux (*Sh*).

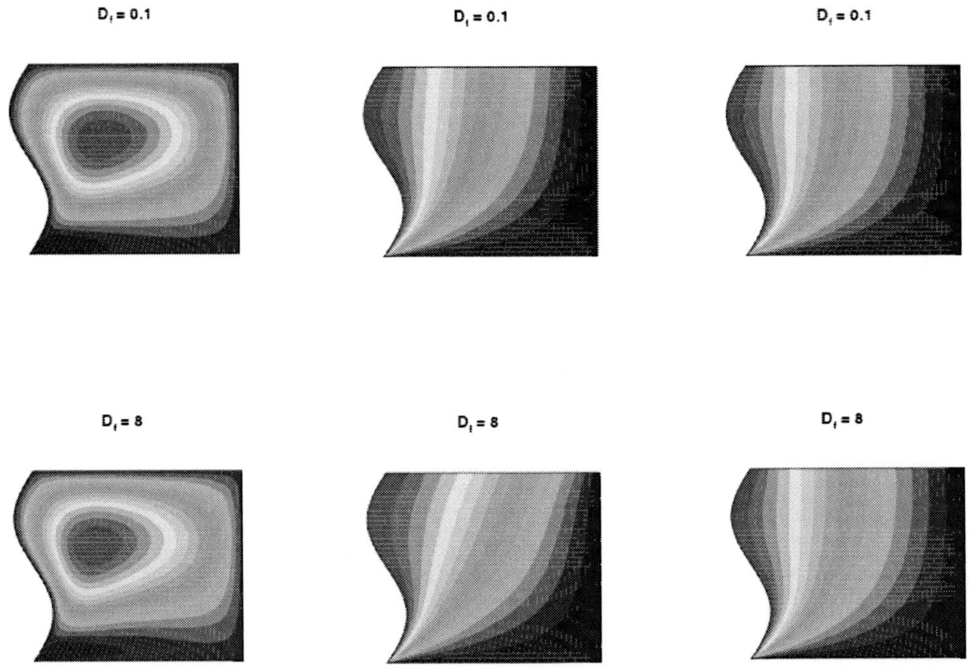

Figure 10. Influence of Dufour effect on flow, temperature and concentration fields traced as streamlines, isotherms and iso-concentration contours respectively.

In Figures 11-12 variation in the cumulative heat and mass fluxes along the wavy wall with increasing values of *N* are presented for D_f, S_r = 0.5, *Le, N, B* = 1, *a* = 0.1 and $1 \leq N \leq 6$. Both the fluxes decrease with increasing values of *N*. The nearly smooth stair case nature in the heat/mass flux plots is due to periodic boost to the thermal/mass related buoyancy forces along the wavy wall. The positive slope of the tangent to the wavy surface indicates the presence of favorable additional buoyancy forces, as they are in the upward direction like those of gravitational buoyancy forces. While one moves from crest to trough the slope is positive and hence there is a raise in the heat/mass flux corresponding to this region. To further analyze the net fall in the heat/mass fluxes with increasing N the corresponding flow, temperature and concentration fields are tracked through streamlines, isotherms and iso-concentration contours in the plots of Figures 13-15. Streamline plots in Figure 13 depict the manifestation of complex multi-cellular circulation pattern, which can go onto hinder; the heat/mass transfer into the core of the domain. Isotherm/Iso-concentration patterns depict a loss in heat/mass flux favoring thermal/concentration boundary layer and a localization of heat/mass with increasing values of N.

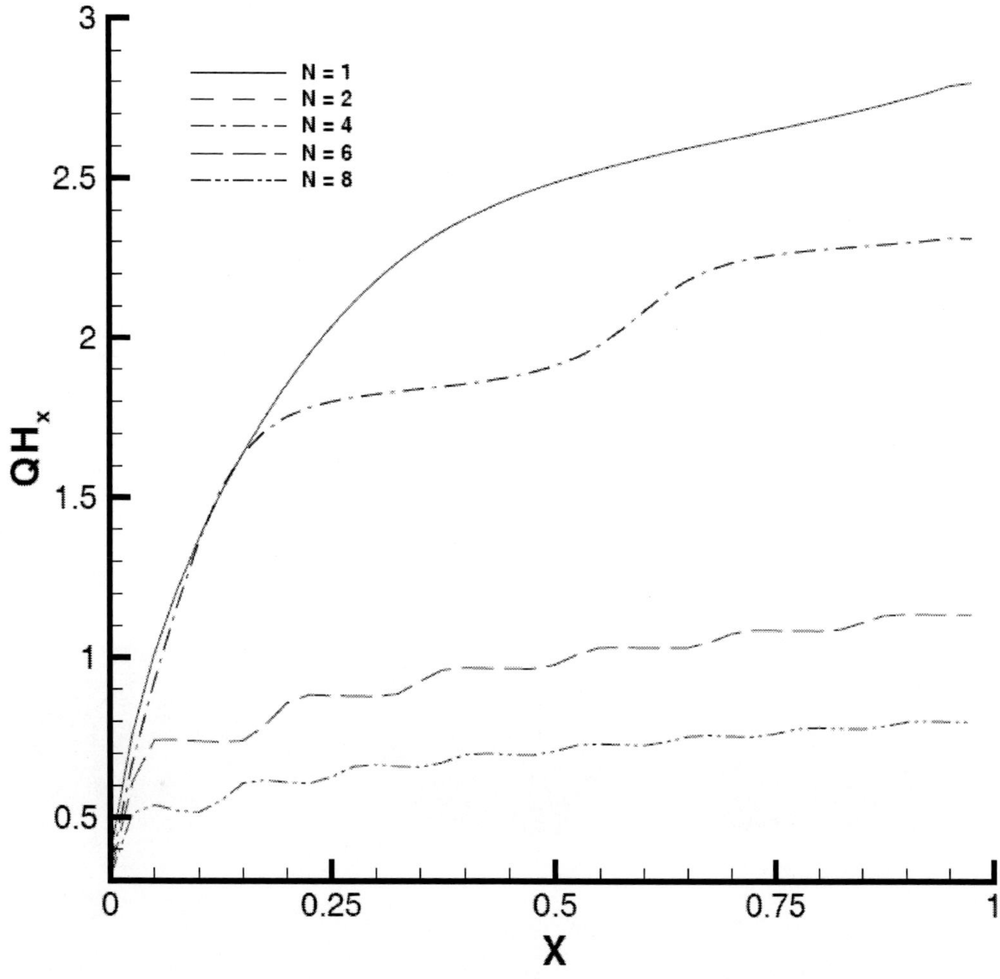

Figure 11. Cumulative Global Heat Flux (QH_X) along the wavy wall with increasing level of corrugation along the wavy wall.

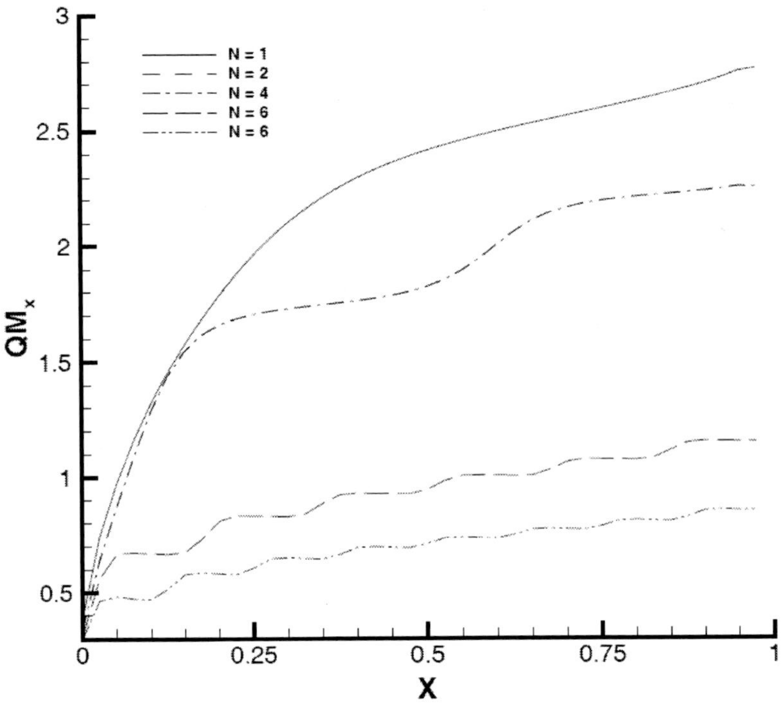

Figure 12. Cumulative Mass Flux (QM_X) along the wavy wall with increasing levels of corrugation on the wavy wall.

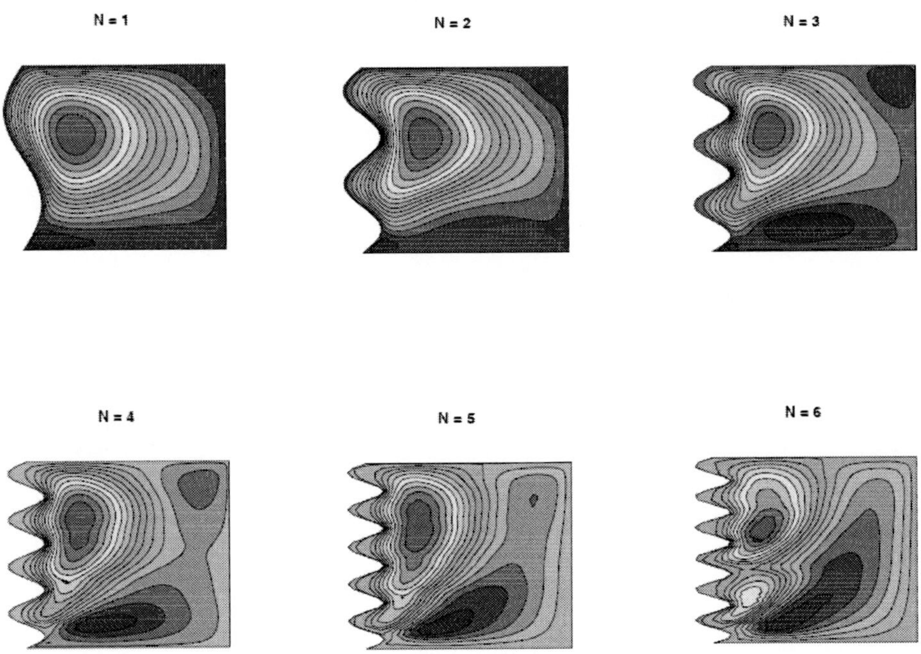

Figure 13. Influence of increasing levels of corrugation on the flow domain traced in the form of streamlines.

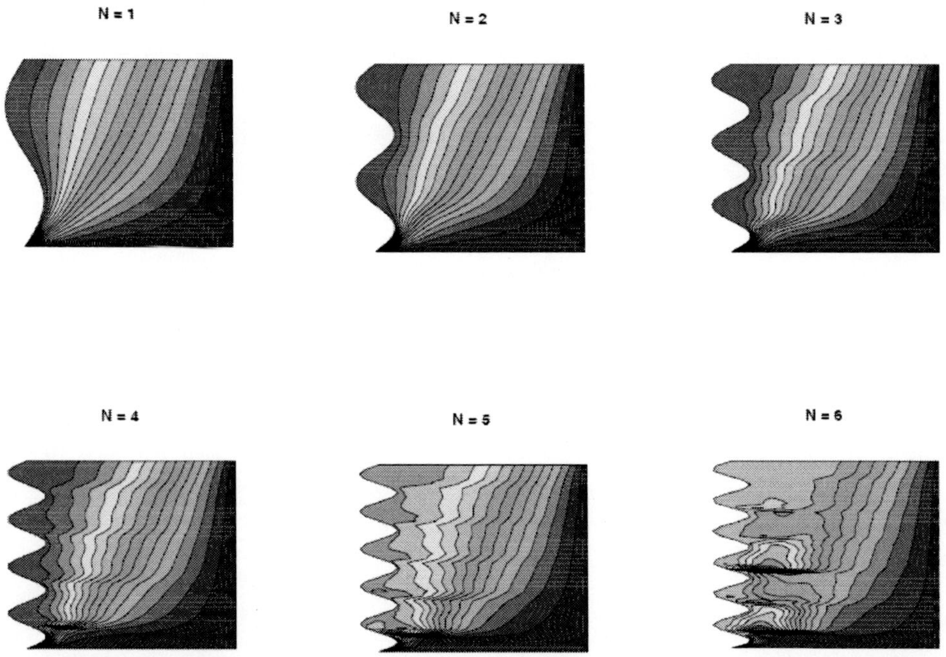

Figure 14. Influence of increasing levels of corrugation on the wavy wall on the temperature field traced as isotherms.

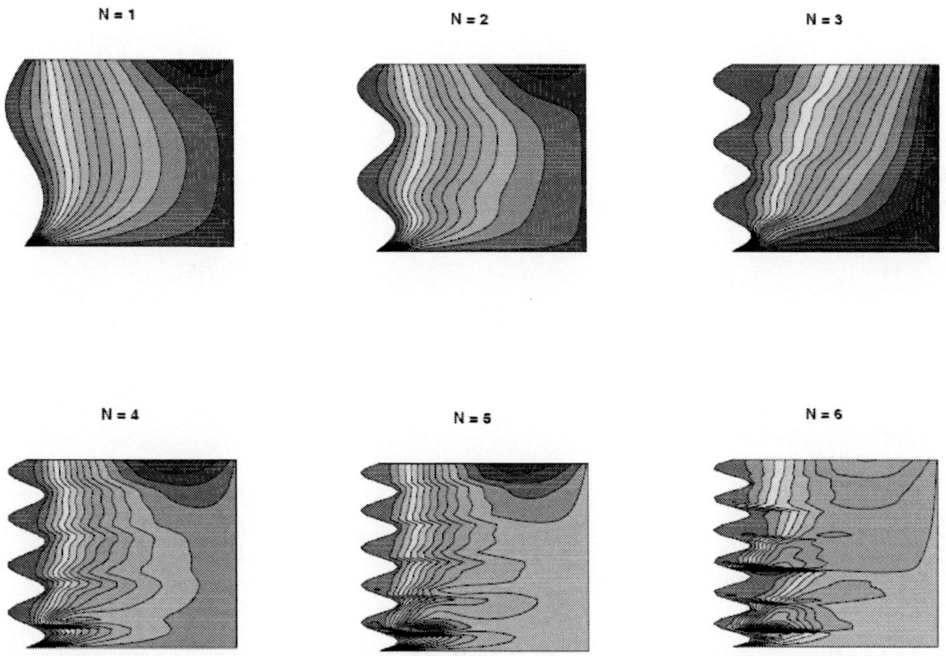

Figure 15. Influence of increasing levels of corrugation on the wavy wall on the concentration field traced in the form of iso-concentration contours.

Finally, the influence of *Le* and *B* on heat/mass fluxes in the presence of both Soret and Dufour effects for $D_f, S_r = 0.5$, $N = 1$, $a = 0.1, 0.1 \leq Le \leq 4$, $-2 \leq B \leq 2$ are analyzed. In Figure 16 the variation of *Nu* and *Sh* with *Le* and *B* are presented. While *Nu* is seen to increase with increasing either *Le* or *B*, *Sh* is seen to decrease. This is exactly contrary to what is observed in the absence of Soret and Dufour effects. So to further analyze the influence of *Le* on the distribution of the field variables streamlines, isotherms and iso-concentration plots traced and presented in Figures 17-19. While the isotherms get shifted from vertical to diagonal orientation, iso-concentration contours get into two localized patterns. Streamlines depict the development of complex multiple circulation zones with increasing Le. All the field variables are sensitive the magnitude of the variation in the ratio of thermal to mass diffusivities. Any increase in thermal diffusivity co-efficient and any decrease in mass diffusivity co-efficient are seen to enhance/reduce heat/mass fluxes into the domain. On increasing *B,* in the presence of Soret and Dufour effects, while *Nu* is marginally increasing *Sh* is seen to marginally decreasing. With varying *B* as opposing thermal and species buoyancy forces begin to aid each other the flow pattern is seen to get complex with horizontal circulation patterns.

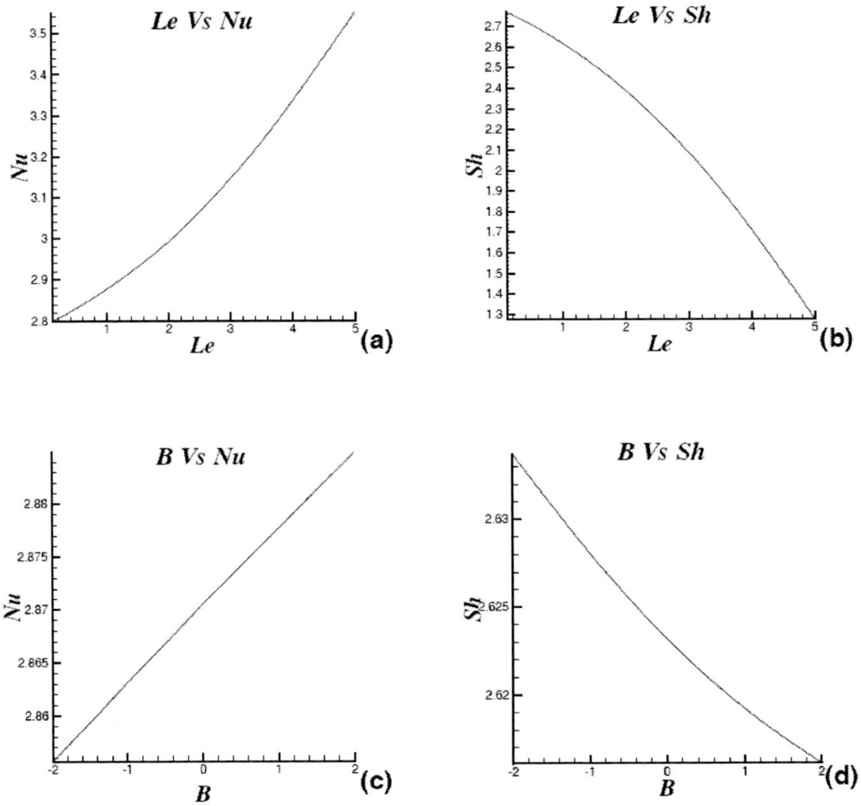

Figure 16. Influence of *Le* and *B* on Global Heat Flux (*Nu*) and Global Mass Flux (*Sh*).

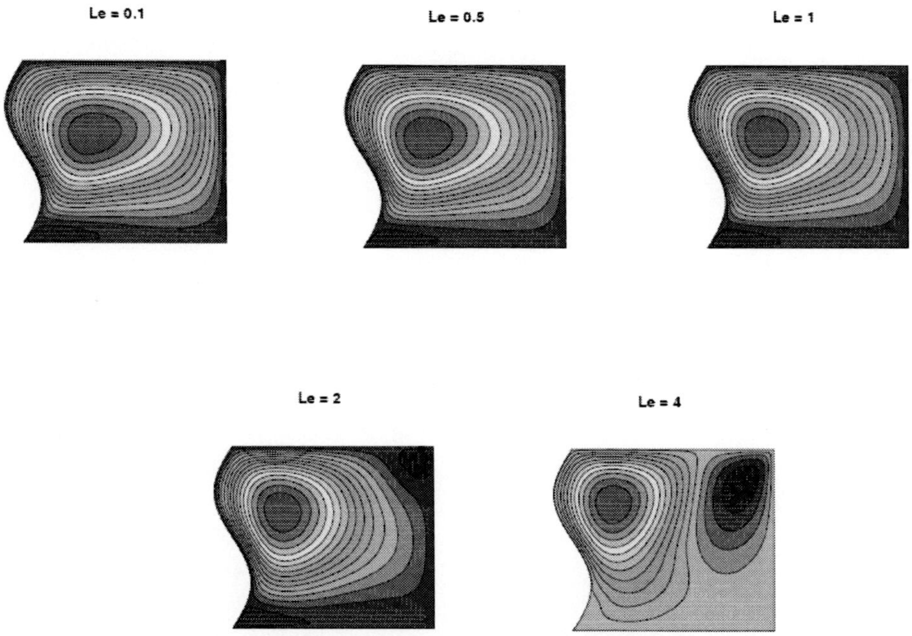

Figure 17. Influence of *Le* on flow field traced in the form of streamlines.

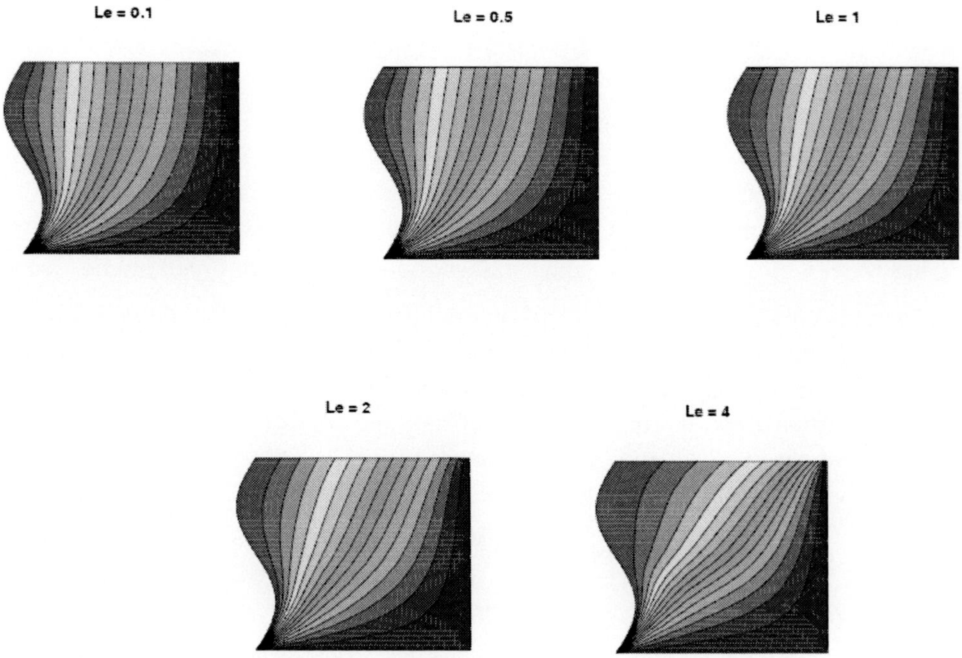

Figure 18. Influence of *Le* on temperature field traced in the form of isotherms.

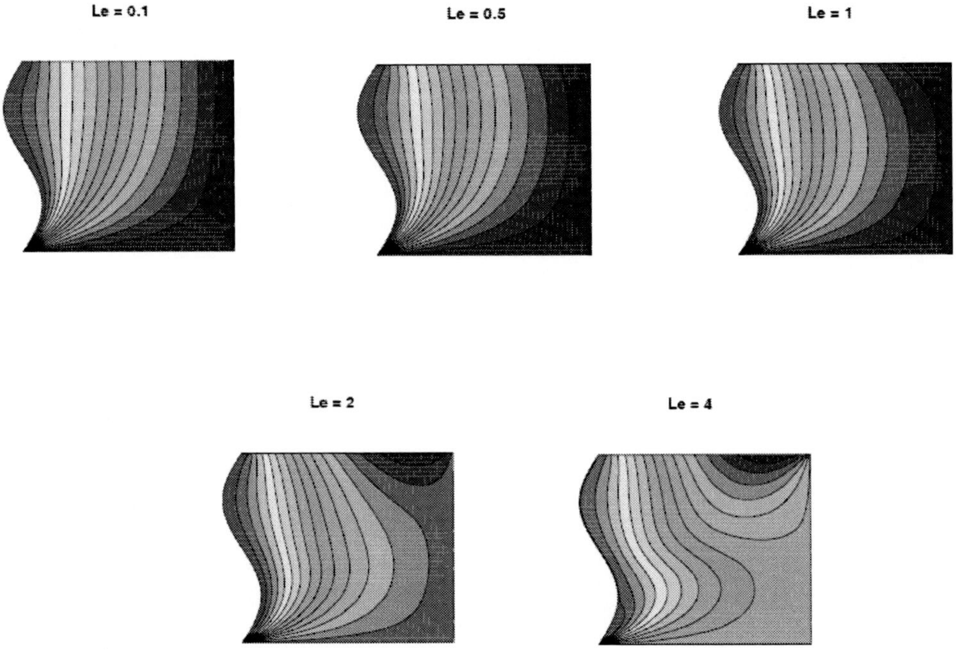

Figure 19. Influence of *Le* on concentration field traced in the form of iso-concentration contours.

CONCLUSIONS

A numerical study based on finite element computation has been carried out to investigate the influence of Soret and Dufour effects on double diffusive natural convection induced by an isothermal wavy vertical wall in a fluid saturated isotropic corrugated porous enclosure. In the presence of Soret and Dufour effects while Nusselt Number increases with increasing values of S_r, D_f, Le and B Sherwood Number is found to be decreasing. However, in the presence of Soret and Dufour effects an increase in N and thereby the surface roughness, weakens both the heat and mass fluxes into the domain. Interesting features like a diagonal shift in the isotherm patterns, the development of multiple localized iso-concentration patterns and the manifestation of multiple complex circulation patterns in the flow domain are observed. Overall at small values of Ra, the influence of Soret effect on the double-diffusive process is more prominent than that of Dufour effect.

REFERENCES

[1] A. Bejan & K.R.Khair, *IJHMT,* 28(5), 909, 1985.
[2] A. Nakayama, M. A. Hossain, *IJHMT,* 38 (4), 761, 1995.
[3] D. Angirasa, G.P.Peterson, I.Pop, *IJHMT*, 40(12), 2755, 1997; NHT: Part A, 34, 1189, 1997.
[4] C. Y. Cheng, *Int. Comm. HMT.* 27(8), 1143, 2000.
[5] B.V.Rathish Kumar, P. Singh, V. J. Bansod, *NHT:* Part A, 41, 421, 2002.

[6] B.V.Rathish Kumar, P. Singh, V. J. Bansod , *JPM* 5(1), 57, 2002.
[7] E. R.G. Eckert and R. M. Drake, 1972, McGraw Hill, NY.
[8] G. Zimmerman and U. Muller, 1992, *IJHMT* 35, 2245-2256.
[9] D.T.Hurle and E. Jakerman, 1989, *JFM.* 447, 667-687.
[10] T. L. Bergman and R. Srinivasan, 1989, *IJHMT,* 32, 679-687.
[11] J. A. Weaver and R. Viskanta, 1991, *IJHMT,* 34, 3121-3133.
[12] L. B. Benano-Melly, J.P. Caltagirone, B. Faissat, F. Montel, and P. Costesque, 2001, *IJHMT,* 44, 1285.
[13] M. Anghel, H.S. Thakar, and I. Pop, 2000, Studia. Universitatis Babes-Bolyai, *Mathematica,* Vol. XLV, P.11.
[14] A. Postelnicu, 2004, *IJHMT,* 47, 1467-1472.
[15] O. Sovran, M.C. Charrier-Mojtabi, A. Mojtabi, 2001, *CR Acad, Sci.* Paris, 239, p. 287.
[16] M.K.Partha, Ph.D.Thesis, IIT-Kgp, India, 2006.
[17] D.S. Riley, *IJHMT,* 31, 1988, 2365-2380.
[18] D.A.S. Rees and I. Pop, *Fluid Dynamics Research,* 14, 1994, 151-166.
[19] D. A. S. Rees and I. Pop, *ASME J. of Ht. Trans.,* Vol. 116, PP. 505-508, 1994.
[20] B.V. Rathish Kumar, P. Singh and P.V.S.N. Murthy, *ASME IJHMT.* 199, 1997, 848-851.
[21] B.V. Rathish Kumar, P.V.S.N. Murthy and P.Singh, *IJNMF,* 28, 1998,633-661
[22] P. V. S. N. Murthy, B. V. Rathish Kumar and P. Singh, N*um. Heat. Transfer,* Part A, 31, 207-221, 1997.
[23] B. V. Rathish Kumar and Shalini, *Applied Mathematics & Computation,* Vol. 17(1), pages 180-222, Dec 2005.
[24] B. V. Rathish Kumar and Shalini, *Journal of Porous Media,* vol. 7, issue 4, pp. 13-30, 2004.
[25] C. Beckermann, R. Viskanta, and S. Ramadhyani., *Numerical Heat Transfer - A*, 10:557–570, 1986.
[26] G. S. Shiralkar, M. Haajizadeh, and C. L. Tien,. *Numerical Heat Transfer - A*, 6:223–234, 1983
[27] O. V. Trevisan and A. Bejan., *Int. J. Heat Mass Transfer*, 28:1597–1611, 1985
[28] K. L. Walker and G. M. Homsy. *J. Fluid Mech.*,87: 449–474, 1978

INDEX

2

3

A

B

C

D

E

F

M

Q

R

S

T

U

V